10 Georgia Milestones Grade 6 Math Practice Tests

The Ultimate Test Prep Collection with Answer Explanations

Dr. A. Nazari

10 Practice Tests

The Grand Championship Collection

Welcome, future Math Champion!

*You hold the **ultimate collection** —
ten full-length practice tests designed to take you
from first attempt to **complete mastery**.*

Conquer every Grade 6 topic

Build unshakeable confidence

Rise from Bronze to Gold to Champion

Arrive at test day fully prepared

The championship begins now.

66 *Ten tests may seem like
a marathon, but champi-
ons are made one step at a
time. Trust the process!* **99**

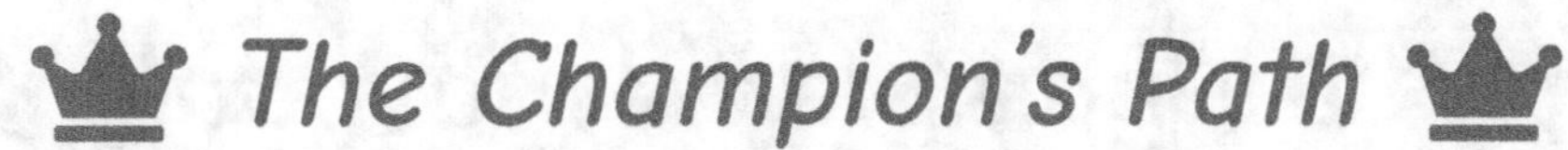

The Champion's Path

Your 4-phase journey from Bronze to Champion

Bronze Round (Tests 1–3)

Your warm-up matches. Take these **untimed** to learn the format and set your baseline. Read the answer explanations after each test — this is where you build your foundation.

Silver Round (Tests 4–6)

Set a timer for **75 minutes**. Focus on the topics that tripped you up in Bronze. Practice showing your work on every problem. Your accuracy should be climbing.

Gold Round (Tests 7–9)

Full timed conditions (**60 minutes**). Simulate the real exam environment. Review only the questions you missed — targeted practice is the key to gold.

Championship Final (Test 10)

Your final match. Full exam conditions — timed, quiet, no breaks. This is your victory lap. Show yourself how far you've come!

Your Championship Kit

- **10 Full-Length Practice Tests** — every Grade 6 topic
- **Formula Reference Sheet**
- **Complete Answer Key** with explanations
- **Championship Scoreboard** to track your rise

Champion's Tip: Space your tests 2–3 days apart. Use the days in between for targeted review. By Test 10, you'll be amazed at your transformation.

♔ The Champion's Playbook ♔

Seven rules that separate champions from the rest

I   **Read every question twice.** *The first read tells you the topic. The second tells you exactly what to solve for. Champions never skim.*

II   **Mark the clues.** *Circle key numbers, underline the question, and cross out information that's just there to distract you.*

III   **Choose your strategy.** *Before touching pencil to paper, decide: Am I setting up a ratio? Solving an equation? Finding area? Name the approach.*

IV   **Solve, then match.** *For multiple choice — work the problem on scratch paper first, then find your answer among the choices.*

V   **Eliminate and conquer.** *Cross out obviously wrong answers. If you're left with two, you've already doubled your odds. Make an educated pick.*

VI   **Estimate to verify.** *After solving, ask: "Is this answer reasonable?" A quick mental estimate catches most calculation errors.*

VII   **Leave nothing blank.** *Even a well-reasoned guess is worth more than an empty space. Use partial work to support your answer.*

🕐 Timing Mastery

Tests 1–3: **Untimed** (build foundation)   ❯   Tests 4–6: **75 min** (build speed)   ❯   Tests 7–10: **60 min** (championship conditions)

★ Grade 6 Championship Topics

🏅 Ratios & Proportions   🏅 Integers & Rational Numbers   🏅 Expressions & Equations

🏅 Geometry & Measurement   🏅 Statistics & Data Analysis

> **A true champion isn't someone who never makes mistakes — it's someone who learns from *every single one*. After each test, review your errors carefully. That's where the real growth happens.**

Find more at
ViewMath.com/GA-Grade6

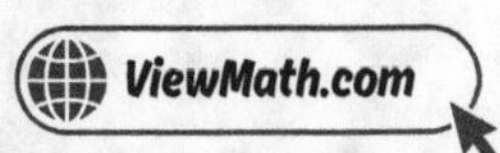

 The Champion's Toolkit

Prepare your workspace before each championship round

	Sharpened Pencils	*Two #2 pencils — champions always have a backup*
	Quality Eraser	*A clean, soft eraser that won't smudge your work*
	Scratch Paper	*Blank paper for calculations, diagrams, and number lines*
	Ruler	*Essential for geometry and coordinate plane questions*
	Timer	*Begin using from the Silver Round onward*
	Quiet Workspace	*A calm, well-lit area free from distractions*

Permitted in Competition

- ✔ Pencil and eraser
- ✔ Scratch paper (provided)
- ✔ Ruler (if specified)
- ✔ Formula reference in this book

Not Permitted

- ✘ Calculators
- ✘ Electronic devices
- ✘ Textbooks or notes
- ✘ Outside help

Find more at
ViewMath.com/GA-Grade6

Formula Reference Sheet

◢ Area Formulas

Rectangle	$A = l \times w$
Parallelogram	$A = b \times h$
Triangle	$A = \dfrac{1}{2} \times b \times h$
Trapezoid	$A = \dfrac{1}{2}(b_1 + b_2) \times h$

▣ Volume

Rectangular Prism $V = l \times w \times h$

▣ Surface Area

Find the area of each face, then add them all up.

Rectangular Prism:

$SA = 2lw + 2lh + 2wh$

↓ Order of Operations

P Parentheses first

E Exponents

M/D Multiply & Divide (left to right)

A/S Add & Subtract (left to right)

% Ratios & Percents

Ratio: $a : b$ or $\dfrac{a}{b}$

Unit rate: amount per 1 unit

Percent: a ratio out of 100

Part = Percent × Whole

⚖ Integers & Absolute Value

Integers:

$$\ldots, -3, -2, -1, 0, 1, 2, 3, \ldots$$

$|-5| = 5 \quad |5| = 5$

Absolute value = distance from 0

X¹ Expressions & Equations

Exponent: $3^4 = 3 \times 3 \times 3 \times 3 = 81$

Variable: a letter that stands for a number

Equation: two expressions joined by $=$

Inequality: uses $<, >, \leq, \geq$

⊕ Coordinate Plane

Ordered pair: (x, y)

x-axis: horizontal y-axis: vertical

Origin: $(0, 0)$

Four quadrants (I, II, III, IV)

▦ Statistics

Mean: sum of values ÷ count

Median: middle value (sorted)

Range: max − min

🏆 Championship Scoreboard 🏆

Track your rise through every championship round

Champion's Name: ______________________________

🏆 Round	🏅 Tier	📅 Date	⭐ Score	😊 Rating
1	Bronze		___ / ___	😊
2	Bronze		___ / ___	😊
3	Bronze		___ / ___	😊
4	Silver		___ / ___	😊
5	Silver		___ / ___	😊
6	Silver		___ / ___	😊
7	Gold		___ / ___	😊
8	Gold		___ / ___	😊
9	Gold		___ / ___	😊
10	👑		___ / ___	😊

X

My strongest topics (where I consistently score well):

____

____

Topics I improved on the most from Bronze to Gold:

____

____

My score trend (Bronze avg → Gold avg → Championship):

____

____

One strategy that helped me improve the most:

____

____

My confidence level for the real test (1–10): _________ / 10

Find more at
ViewMath.com/GA-Grade6

Continue Learning at
ViewMath Academy!

For Parents, Teachers & Students

Great job on the practice tests! Want to keep improving? ViewMath Academy is your **free online companion** to this book.

- **Score Analyzer** — Enter your answers and instantly see which topics need more practice

- **Interactive Lessons** — Review the concepts behind each question with clear explanations

- **Adaptive Quizzes** — Practice your weak topics with questions that match your level

- **Progress Tracking** — See your mastery grow across all Grade 6 math topics

- **Personalized Dashboard** — A learning plan tailored just for you

Scan to visit ViewMath Academy

ViewMath.com/GA-Grade6

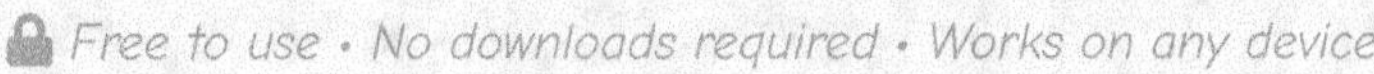 Free to use · No downloads required · Works on any device 

★ Table of Contents ★

Here's what we'll explore together!

 Let's learn and have fun!

Practice Test 1

30 Questions

✏️ Before You Start ✏️

- ✔ **Read each question carefully** before choosing your answer.
- ✔ **Show your work** on scratch paper when you need to.
- ✔ **Skip hard questions** and come back to them later.
- ✔ **Check your answers** when you're done.
- ✔ **Take your time** — there's no rush!

⭐ You've Got This! ⭐

Do your best and show what you know!

1. A garden has flowers and vegetables in a ratio of 4 : 1. There are 20 flowers. How many vegetables are there?

Your Answer:

2. The graph below shows the distance traveled by a delivery truck over time.

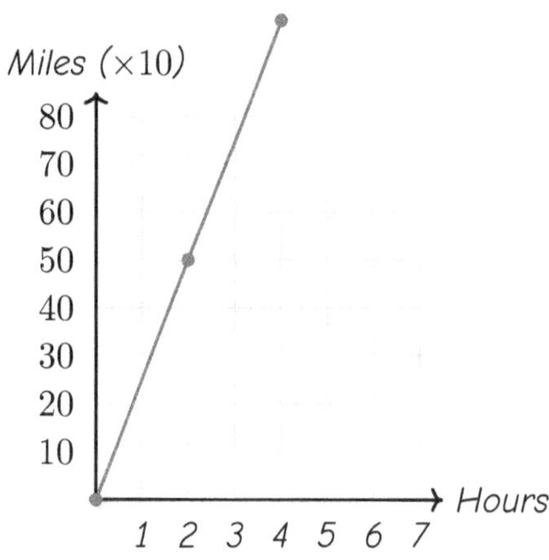

What is the truck's unit rate in miles per hour?

(A) 50 miles per hour

(B) 20 miles per hour

(C) 25 miles per hour

(D) 10 miles per hour

3. The double number line below shows equivalent ratios of scoops of mix to cups of water.

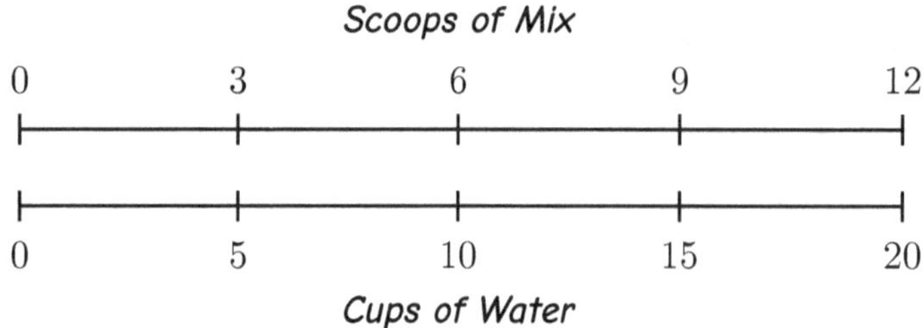

How many cups of water go with 9 scoops of mix?

(A) 10

(B) 12

(C) 15

(D) 18

Find more at
ViewMath.com/GA-Grade6

4. A graph of a ratio relationship passes through $(0,0)$ and $(5,20)$. What is the unit rate?

(A) 5

(B) 20

(C) 4

(D) 25

---

5. The double number line below shows a map scale.

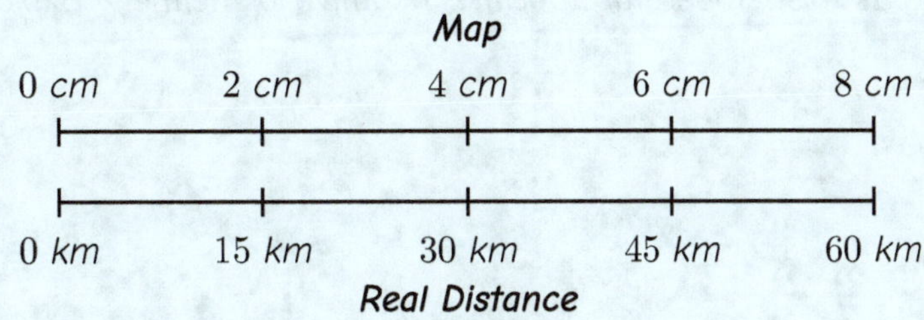

Two cities are 5 cm apart on the map. What is the real distance?

(A) 25 km

(B) 30 km

(C) 37.5 km

(D) 35 km

---

6. A bottle holds 750 mL. How many full bottles can you fill from a 6-liter jug?

Your Answer:

---

7. The number line below shows equal jumps from 0 to $\frac{5}{6}$.

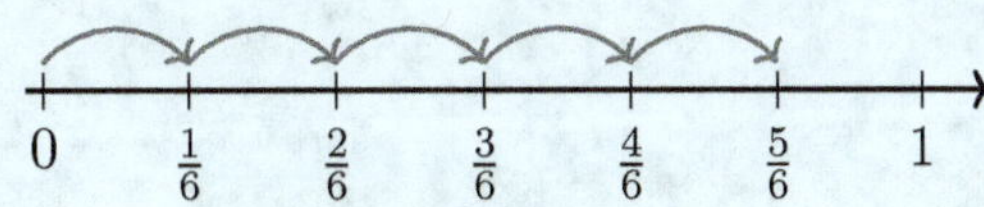

Which division problem does the number line represent?

(A) $\dfrac{5}{6} \div \dfrac{1}{6} = 5$

(B) $\dfrac{1}{6} \div \dfrac{5}{6} = 5$

(C) $\dfrac{5}{6} \times \dfrac{1}{6} = \dfrac{5}{36}$

(D) $\dfrac{5}{6} \div 5 = \dfrac{1}{6}$

8. Liam reads 1,344 pages in 8 weeks, reading the same number of pages each week. How many pages does he read per week?

(A) 158

(B) 178

(C) 186

(D) 168

9. A thermometer reads $0°C$. Is this temperature positive, negative, or neither? Explain.

Your Answer:

10. Order these numbers from least to greatest: $\frac{1}{2}, -\frac{3}{4}, 0.2, -1.5, \frac{5}{4}$.

Your Answer:

11. A store charges the prices shown below. Which expression gives the total cost for s sandwiches and d drinks?

Sandwich	Drink
$6 each	$2 each

(A) $6 + 2$

(B) $6s + 2d$

(C) $8sd$

(D) $2s + 6d$

Find more at
ViewMath.com/GA-Grade6

12. A student made a factor tree for the expression $4(x+3)$, shown below. Fill in the missing pieces labeled A and B.

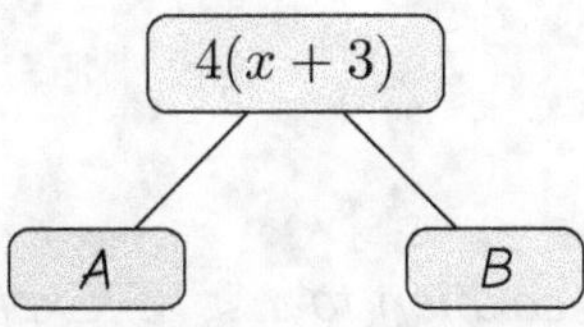

Your Answer:

13. Evaluate $3(p+2) - p$ when $p = 8$.

Your Answer:

14. Which expression is equivalent to $3(x+5)$?

(A) $3x + 5$

(B) $3x + 15$

(C) $8x$

(D) $3x + 8$

15. Write a situation that could be modeled by the expression $5n + 3$.

Your Answer:

16. Solve: $\dfrac{m}{4} = 7$

(A) $m = 3$

(B) $m = 11$

(C) $m = 28$

(D) $m = 47$

17. *Write a real-world situation that can be described by $x \leq 15$.*

Your Answer

18. *Is $x = 3$ a solution to $x \leq 3$? Is $x = 3$ a solution to $x < 3$? Explain.*

Your Answer

19. *Which table shows a relationship where y does NOT depend on x?*

(A) $x : 1, 2, 3 \quad y : 3, 6, 9$

(B) $x : 1, 2, 3 \quad y : 7, 7, 7$

(C) $x : 1, 2, 3 \quad y : 2, 4, 6$

(D) $x : 1, 2, 3 \quad y : 5, 10, 15$

20. *A composite figure is made of a rectangle (8 cm by 5 cm) attached to a triangle with base 8 cm and height 3 cm. What is the total area?*

(A) 52 cm^2

(B) 40 cm^2

(C) 55 cm^2

(D) 64 cm^2

21. Volume is measured in which type of units?

(A) Square units (cm^2)

(B) Linear units (cm)

(C) Cubic units (cm^3)

(D) No units are needed

22. Points $A(0, 4)$ and $B(0, -5)$ are connected by a line segment. What is the length of $\overline{AB}$?

(A) 1 unit

(B) 5 units

(C) 4 units

(D) 9 units

23. A right triangle has vertices $(2, 1)$, $(2, 7)$, and $(8, 1)$. What is the area?

(A) 36 square units

(B) 18 square units

(C) 12 square units

(D) 24 square units

24. A cube has edge length 3 cm. What is the surface area?

(A) $9\ cm^2$

(B) $27\ cm^2$

(C) $36\ cm^2$

(D) $54\ cm^2$

25. When data is "skewed to the right," what does that mean?

(A) Most values are on the left, with a tail stretching to the right.

(B) Most values are on the right, with a tail stretching to the left.

(C) The data is symmetric.

(D) All values are equal.

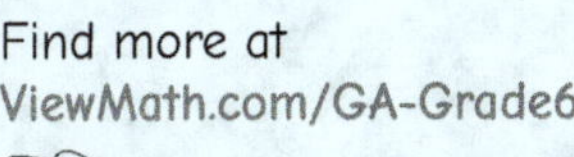
Find more at
ViewMath.com/GA-Grade6

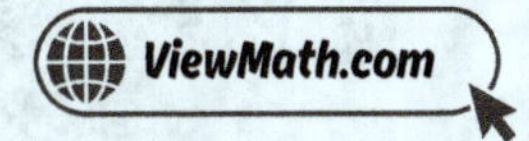

26. A frequency table shows how many books students read last month:

Books	Students
1	3
2	5
3	4
4	2
5	1

What is the median number of books?

(A) 2

(B) 2.5

(C) 3

(D) 4

27. Data (in order): $3, 5, 7, 8, 10, 12, 14$. What is the IQR?

(A) 5

(B) 7

(C) 9

(D) 11

28. A box plot shows: $min = 25$, $Q1 = 35$, $median = 50$, $Q3 = 65$, $max = 80$. Find the range and IQR.

29. Two classes both have a mean of 75. Class A: $MAD = 3$. Class B: $MAD = 9$. A student from each class is picked at random. Whose score is more likely to be close to 75?

(A) The student from Class A

(B) The student from Class B

(C) Both equally likely

(D) Cannot determine

30. A spinner has 8 equal sections. Three sections are blue and five sections are red. What is the probability of **not** landing on blue? Write your answer as a decimal.

Your Answer:

Find more at
ViewMath.com/GA-Grade6

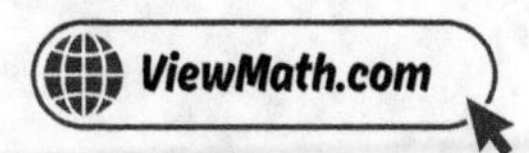

 End of Practice Test 1 

Great job finishing the test!

 My Score

I got _____________ out of 30 questions right.

*Check your answers in the **Answer Key** at the back of the book.*

💡 *Review any questions you missed. That's how we learn!*

📊 Check Your Score Online!

Visit **ViewMath Academy** to enter your answers and see which topics you need to review. You can also explore lessons, take quizzes, track your scores, and save your progress!

viewmath.com/score/6.1.GA.16

*Or go to **viewmath.com/score** and enter code: 6.1.GA.16*

2

Practice Test 2

 30 Questions

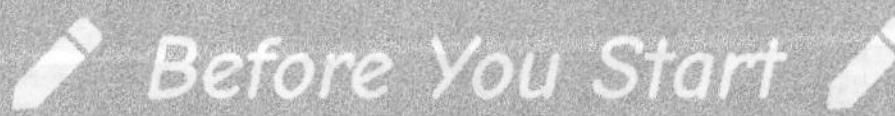 Before You Start

- ✓ **Read each question carefully** before choosing your answer.
- ✓ **Show your work** on scratch paper when you need to.
- ✓ **Skip hard questions** and come back to them later.
- ✓ **Check your answers** when you're done.
- ✓ **Take your time** — there's no rush!

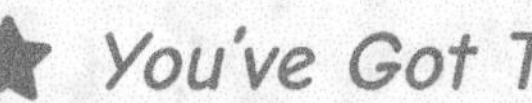 You've Got This! 

Do your best and show what you know!

1. Look at the tape diagram below.

Boys: ▢▢▢▢
Girls: ▢▢▢▢▢▢

There are 24 girls. How many boys are there?

(A) 4   (B) 12

(C) 16   (D) 20

2. The table shows the earnings of two babysitters.

	Hours Worked	Earnings
Lily	5	$60
Noah	4	$52

Part A: Find each babysitter's unit rate (dollars per hour).

Part B: Who earns more per hour? How much more?

Your Answer

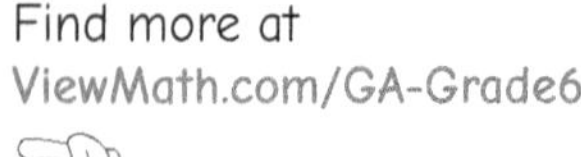
Find more at
ViewMath.com/GA-Grade6

3. The graph below shows points that represent equivalent ratios of flour to sugar in a recipe.

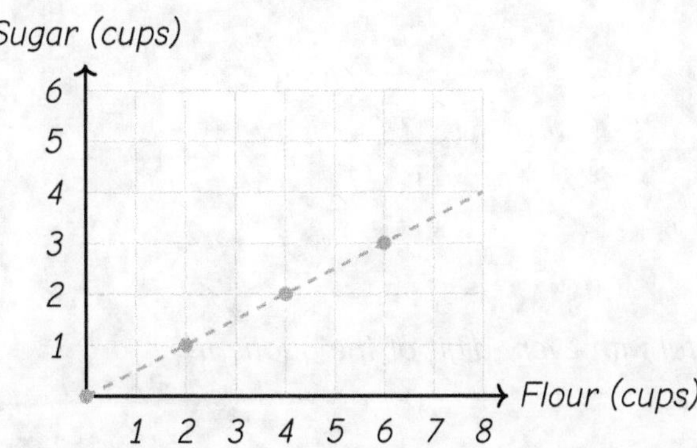

Part A: What is the ratio of flour to sugar?

Part B: If you need 8 cups of flour, how many cups of sugar do you need?

Your Answer:

4. The points $(2, 5)$ and $(4, 10)$ are on a graph. What is the ratio $x : y$?

(A) $5 : 2$ (B) $1 : 2$

(C) $2 : 5$ (D) $4 : 5$

5. A pool fills at a rate of 12 gallons per minute. How long does it take to fill 360 gallons?

(A) 25 minutes   (B) 30 minutes

(C) 36 minutes   (D) 40 minutes

6. A race is 10 kilometers. How many centimeters is that?

(A) 100,000   (B) 10,000

(C) 1,000,000   (D) 1,000

7. What is $\dfrac{3}{5} \div \dfrac{1}{5}$?

   (A) $\dfrac{3}{25}$ (B) 3

   (C) $\dfrac{1}{3}$ (D) 15

8. Compute $4{,}515 \div 15$. Be careful with every digit of the quotient.

Your Answer

9. Maria has \$0 in her bank account. What does this mean?

   (A) She owes money.   (B) She has some savings.

   (C) She has no money and no debt.   (D) She has negative money.

10. What rational number is exactly halfway between $\dfrac{1}{4}$ and $\dfrac{3}{4}$ on a number line?

Your Answer

11. A baker makes c cupcakes. She packs them equally into 6 boxes. Which expression shows the number in each box?

   (A) $c - 6$ (B) $6c$

   (C) $c \div 6$ (D) $c + 6$

12. In the expression $5(y + 8)$, name the two factors.

Your Answer

Find more at
ViewMath.com/GA-Grade6

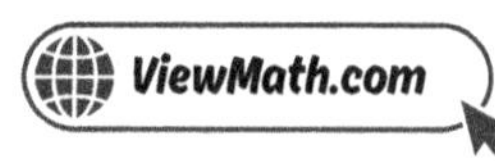

13. The triangle below has a base of $b = 10$ cm and a height of $h = 6$ cm. Use the formula $A = \frac{1}{2} \times b \times h$ to find its area.

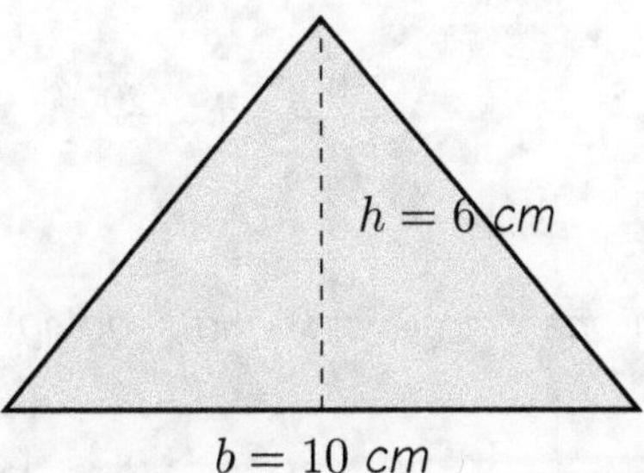

Your Answer:

14. Simplify: $3a + 7 + 5a - 2a + 3$

Your Answer:

15. A movie ticket costs \$9. Your group also shares a \$6 bucket of popcorn. Which expression represents the total cost for p people?

(A) $9p + 6$

(B) $9 + 6p$

(C) $15p$

(D) $9p - 6$

16. Solve: $k + 15 = 32$

(A) $k = 47$

(B) $k = 17$

(C) $k = 27$

(D) $k = 2$

17. Which inequality represents "a number y is fewer than 20"?

(A) $y > 20$

(B) $y \geq 20$

(C) $y < 20$

(D) $y \leq 20$

18. Which inequality has a graph with an open circle at 0 and shading to the right?

(A) $x \leq 0$

(B) $x \geq 0$

(C) $x > 0$

(D) $x < 0$

19. Identify the independent and dependent variables: A runner jogs 6 miles per hour. The distance d depends on how many hours h the runner jogs.

Your Answer

20. A parallelogram has base 12 m and a slant side of 8 m. The height is 6 m. What is the area?

(A) $96\ m^2$

(B) $72\ m^2$

(C) $48\ m^2$

(D) $36\ m^2$

21. A storage container is 5 m long, 3 m wide, and 2 m tall. It is half-full of sand. How many cubic meters of sand are in the container?

(A) $30\ m^3$

(B) $15\ m^3$

(C) $10\ m^3$

(D) $60\ m^3$

22. A rectangle has vertices $(-6, 2)$, $(2, 2)$, $(2, -3)$, $(-6, -3)$. What is the perimeter?

Your Answer

Find more at
ViewMath.com/GA-Grade6

23. To find the area of an irregular polygon on the coordinate plane, you can:

(A) Multiply all the coordinates together.

(B) Break it into rectangles and triangles, then add the areas.

(C) Count only the vertices.

(D) Subtract the perimeter from the largest coordinate.

24. A rectangular prism has dimensions 2 ft by 3 ft by h ft. Its surface area is 62 ft^2. What is h?

Your Answer:

25. Look at the dot plot below.

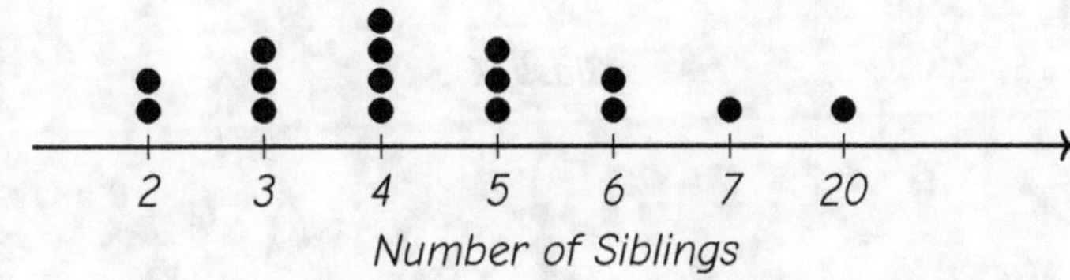

Which value is an outlier?

(A) 2

(B) 4

(C) 7

(D) 20

26. The dot plot below shows daily high temperatures (°F) for 7 days.

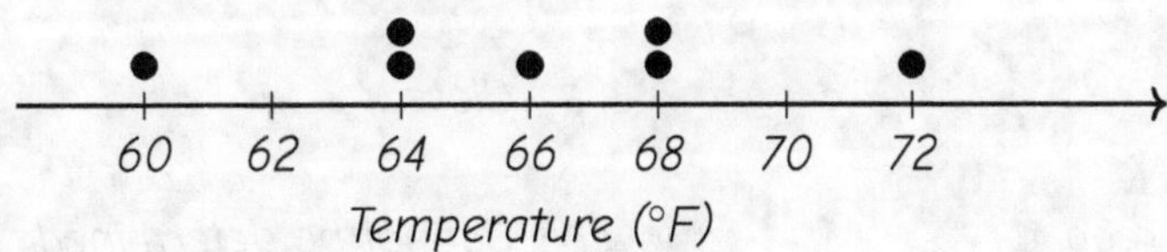

Find the mean and median temperature from the dot plot.

Your Answer:

Find more at
ViewMath.com/GA-Grade6

27. What is the range of the data set $8, 12, 15, 20, 25$?

(A) 8

(B) 13

(C) 17

(D) 25

28. Which data set has the five-number summary: $5, 10, 15, 20, 25$?

(A) $5, 10, 15, 20, 25$

(B) $5, 8, 10, 15, 18, 20, 22, 25$

(C) $5, 7, 10, 12, 15, 18, 20, 23, 25$

(D) Any of these could have that five-number summary.

29. The dot plots below show the number of sit-ups completed by students in two PE classes.

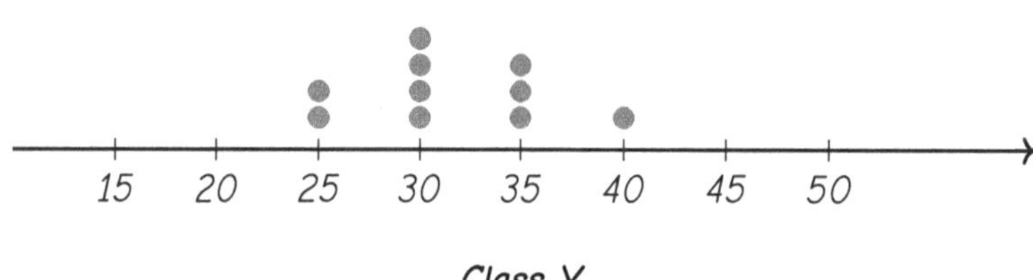

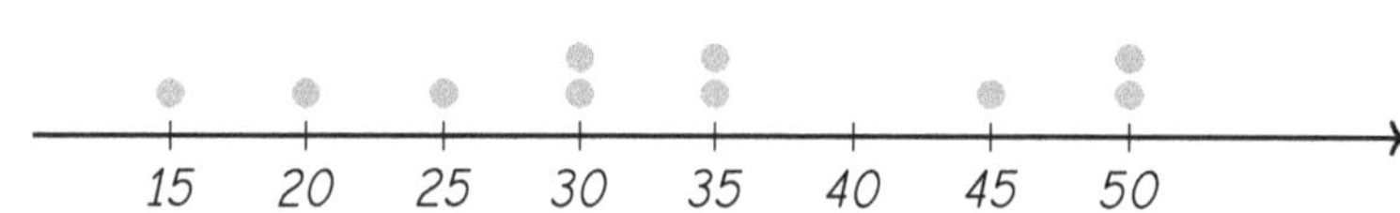

Which class has data that is more spread out?

(A) Class X

(B) Class Y

(C) They have the same spread.

(D) Cannot be determined.

30. The probability scale below shows four events. Which event has a probability closest to 0.75?

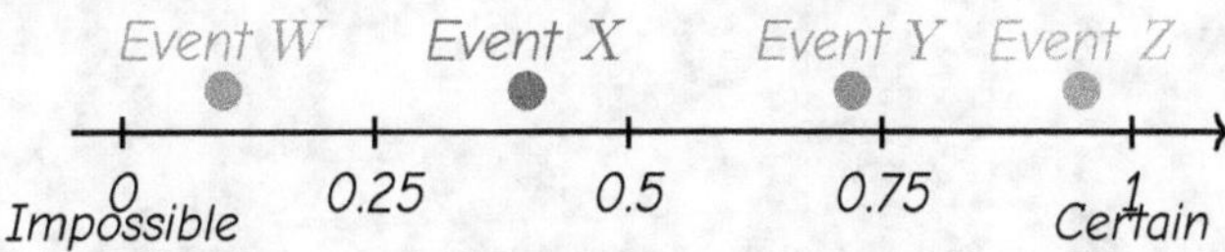

(A) Event W

(B) Event X

(C) Event Y

(D) Event Z

Find more at
ViewMath.com/GA-Grade6

 *End of Practice Test 2* 

Great job finishing the test!

☑ My Score

I got ____________ out of 30 questions right.

Check your answers in the Answer Key at the back of the book.

💡 *Review any questions you missed. That's how we learn!*

📊 Check Your Score Online!

*Visit **ViewMath Academy** to enter your answers and see which topics you need to review. You can also explore lessons, take quizzes, track your scores, and save your progress!*

viewmath.com/score/6.1.GA.17

Or go to viewmath.com/score and enter code: 6.1.GA.17

Practice Test 3

 30 Questions

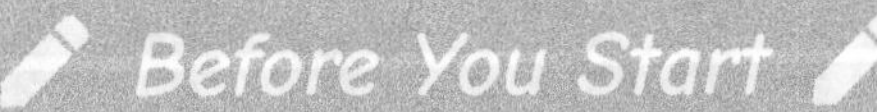

✏️ Before You Start ✏️

- ✔ **Read each question carefully** before choosing your answer.
- ✔ **Show your work** on scratch paper when you need to.
- ✔ **Skip hard questions** and come back to them later.
- ✔ **Check your answers** when you're done.
- ✔ **Take your time** — there's no rush!

 You've Got This! 

Do your best and show what you know!

1. A baker uses 2 eggs **per** 3 cups of flour. Which ratio represents flour to eggs?

   (A) $2:3$ (B) $3:5$

   (C) $3:2$ (D) $2:5$

2. A box of 12 muffins costs \$9. What is the cost per muffin?

   (A) \$1.25   (B) \$0.75

   (C) \$1.33   (D) \$3.00

3. A store sells 7 bananas for \$2. How much do 21 bananas cost?

   (A) \$4   (B) \$6

   (C) \$9   (D) \$14

4. A ratio graph passes through $(6, 2)$. What is the value of y when $x = 15$?

   (A) 3   (B) 5

   (C) 7   (D) 10

5. A store sells 3 T-shirts for \$24. How much do 10 T-shirts cost?

   (A) \$72   (B) \$80

   (C) \$70   (D) \$90

6. Convert 3 miles to feet. (1 mile = 5,280 feet)

   (A) 5,280   (B) 10,560

   (C) 15,840   (D) 21,120

7. What is $\dfrac{1}{3} \div \dfrac{2}{3}$?

(A) $\dfrac{1}{2}$ (B) $\dfrac{2}{9}$

(C) $\dfrac{2}{3}$ (D) 2

8. What is $1{,}575 \div 5$?

(A) 305 (B) $3{,}150$

(C) 315 (D) 351

9. A helicopter is flying at 1,200 feet above sea level. Which integer best describes its altitude?

(A) $-1{,}200$ (B) 0

(C) $1{,}200$ (D) -12

10. What number is exactly halfway between -1 and 0 on a number line?

(A) $-\dfrac{1}{4}$ (B) $\dfrac{1}{2}$

(C) -1.5 (D) $-\dfrac{1}{2}$

11. Sam has x stickers. He gives away 6. Which expression shows how many stickers Sam has now?

(A) $x + 6$ (B) $6x$

(C) $6 - x$ (D) $x - 6$

12. How many terms does the expression $4(a + b)$ have when written in its factored form?

Your Answer:

Find more at
ViewMath.com/GA-Grade6

13. Evaluate $k^2 + k$ when $k = 3$.

    (A) 6   (B) 9

    (C) 12   (D) 15

14. Use the distributive property to expand $8(y + 3)$.

Your Answer

15. In the formula $d = 60t$, what could d and t represent?

    (A) d = days, t = hours   (B) d = distance in miles, t = time in hours

    (C) d = dollars, t = tax rate   (D) d = degrees, t = temperature

16. Solve: $x + 24 = 50$

    (A) $x = 26$ (B) $x = 74$

    (C) $x = 36$ (D) $x = 2$

17. The table below matches phrases to inequality symbols. Which row has an error?

Row	Phrase	Symbol
1	At least	$\geq$
2	No more than	$\leq$
3	Fewer than	$<$
4	At most	$<$

    (A) Row 1   (B) Row 2

    (C) Row 3   (D) Row 4

Find more at
ViewMath.com/GA-Grade6

18. The pool opens when the temperature is more than 75°F. Which describes the graph of this inequality?

(A) Closed circle at 75, shade right

(B) Open circle at 75, shade right

(C) Closed circle at 75, shade left

(D) Open circle at 75, shade left

19. A recipe uses 2 cups of flour for every batch of cookies. If you make b batches, which equation gives the total flour f?

(A) $f = b + 2$

(B) $f = 2b$

(C) $b = 2f$

(D) $f = b \div 2$

20. A park is shaped like a trapezoid with parallel sides of 40 m and 60 m and a height of 30 m. What is the area of the park?

(A) $1{,}800\ m^2$

(B) $1{,}200\ m^2$

(C) $2{,}400\ m^2$

(D) $1{,}500\ m^2$

21. A rectangular prism has length $\frac{3}{4}$ ft, width 2 ft, and height 4 ft. What is the volume?

(A) $6\ ft^3$

(B) $3\ ft^3$

(C) $24\ ft^3$

(D) $\frac{3}{2}\ ft^3$

22. *Find the perimeter of the rectangle shown on the coordinate plane.*

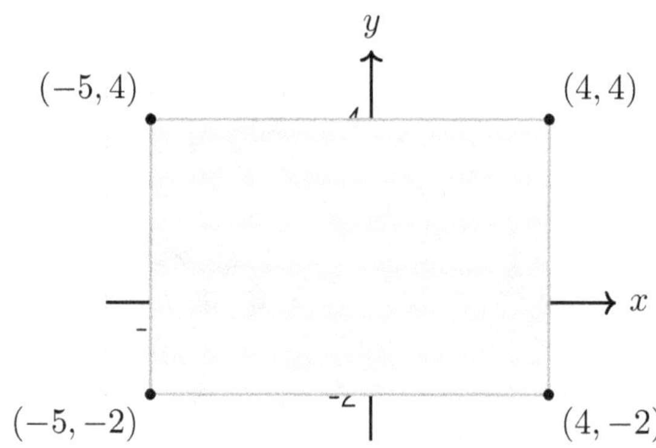

Your Answer:

---

23. *A right triangle has vertices $(1, 1)$, $(1, 9)$, and $(5, 1)$. What is the area?*

(A) 32 square units

(B) 16 square units

(C) 20 square units

(D) 8 square units

---

24. *Mia paints all sides of a wooden block that is 3 ft by 2 ft by 4 ft. She then paints an identical block. How much total area does she paint?*

(A) 52 ft^2

(B) 104 ft^2

(C) 48 ft^2

(D) 26 ft^2

---

25. *Data set A: $20, 21, 22, 23, 24$. Data set B: $10, 15, 22, 29, 34$. Both have a center near 22. What is different?*

(A) Data set A is more spread out.

(B) Data set B is more spread out.

(C) Both have the same spread.

(D) Neither data set has a center.

26. Data: $3, 4, 5, 6, 7, 8, 60$. Find the mean and median. Which better describes a typical value?

Your Answer:

27. Which statement about MAD is true?

(A) MAD measures the middle value of the data.

(B) MAD tells you the average distance of each value from the mean.

(C) MAD is always larger than the range.

(D) MAD equals the range divided by 2.

28. A box plot has min $= 10$, $Q1 = 20$, median $= 30$, $Q3 = 40$, max $= 50$. What is the IQR?

(A) 10

(B) 20

(C) 30

(D) 40

29. Class A: median $= 78$, IQR $= 6$. Class B: median $= 78$, IQR $= 18$. What can you conclude?

(A) Both classes performed identically.

(B) Class B is more consistent.

(C) Class A is more consistent because its IQR is smaller.

(D) Class B has a higher center.

30. A bag has 5 orange balls, 3 purple balls, and 2 yellow balls. What is the probability of **not** drawing a yellow ball?

(A) $\dfrac{2}{10}$

(B) $\dfrac{5}{10}$

(C) $\dfrac{8}{10}$

(D) $\dfrac{3}{10}$

Find more at
ViewMath.com/GA-Grade6

ViewMath.com

 End of Practice Test 3 

Great job finishing the test!

☑ My Score

I got _____________ out of 30 questions right.

*Check your answers in the **Answer Key** at the back of the book.*

💡 *Review any questions you missed. That's how we learn!*

📊 Check Your Score Online!

Visit **ViewMath Academy** to enter your answers and see which topics you need to review. You can also explore lessons, take quizzes, track your scores, and save your progress!

viewmath.com/score/6.1.GA.18

Or go to viewmath.com/score and enter code: 6.1.GA.18

Practice Test 4

 30 Questions

✏️ Before You Start ✏️

- ✔ **Read each question carefully** before choosing your answer.
- ✔ **Show your work** on scratch paper when you need to.
- ✔ **Skip hard questions** and come back to them later.
- ✔ **Check your answers** when you're done.
- ✔ **Take your time** — there's no rush!

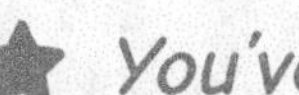 You've Got This! 

Do your best and show what you know!

1. A store sells 3 pencils **for every** 1 eraser. Which ratio represents pencils to erasers?

   (A) $1:3$ (B) $3:1$

   (C) $3:4$ (D) $1:4$

2. A bike travels at a unit rate of 14 miles per hour. How long will it take to travel 49 miles?

   Your Answer:

3. The ratio table below has an error in one row. Which row has the error?

x	y
4	10
8	20
12	25
16	40

   (A) Row 1   (B) Row 2

   (C) Row 3   (D) Row 4

4. On a graph, Line A passes through $(1,6)$ and Line B passes through $(1,4)$. Both go through the origin. Which line represents a faster rate? Explain.

   Your Answer:

5. A car travels 180 miles in 3 hours. At the same rate, how far will it go in 7 hours?

   (A) 360 miles   (B) 420 miles

   (C) 540 miles   (D) 300 miles

Find more at
ViewMath.com/GA-Grade6

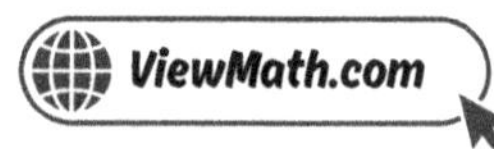

6. *How many cups are in 3 quarts? (1 quart = 4 cups)*

(A) 7

(B) 8

(C) 10

(D) 12

7. *If $\frac{a}{b} \div \frac{c}{d} = \frac{a}{b} \times \frac{?}{?}$, what fraction replaces the question marks?*

(A) $\frac{d}{c}$

(B) $\frac{c}{d}$

(C) $\frac{b}{a}$

(D) $\frac{a}{b}$

8. *A factory produces 5,376 bottles and packs them into cases of 32. How many cases are needed?*

Your Answer

9. *Which of the following numbers is negative?*

(A) 7

(B) 0

(C) -4

(D) 12

10. *The number -1.5 is located between which two integers on a number line?*

(A) -1 and 0

(B) -3 and -2

(C) -2 and -1

(D) 0 and 1

11. *Which phrase matches the expression $\frac{n}{3} + 7$?*

(A) *7 more than a number n divided by 3*

(B) *3 divided by n plus 7*

(C) *7 times n divided by 3*

(D) *n plus 7, divided by 3*

Find more at
ViewMath.com/GA-Grade6

ViewMath.com

12. *Identify the constant(s) in the expression* $3m + 7 - 2m + 4$.

Your Answer

13. *Look at the formula and the rectangle below. What is the perimeter of the rectangle?*

$$w = 3\ cm \qquad \boxed{P = 2l + 2w}$$

$$l = 7\ cm$$

(A) 10 cm

(B) 20 cm

(C) 21 cm

(D) 42 cm

14. *Simplify:* $2(4m + 1) + 3m$

Your Answer

15. *In the formula* $P = 4s$, *the variable* s *represents the side length of a square. What does* P *represent?*

(A) The area of the square

(B) The perimeter of the square

(C) The number of sides

(D) The diagonal of the square

16. The number line below shows a jump. Which equation does this model represent?

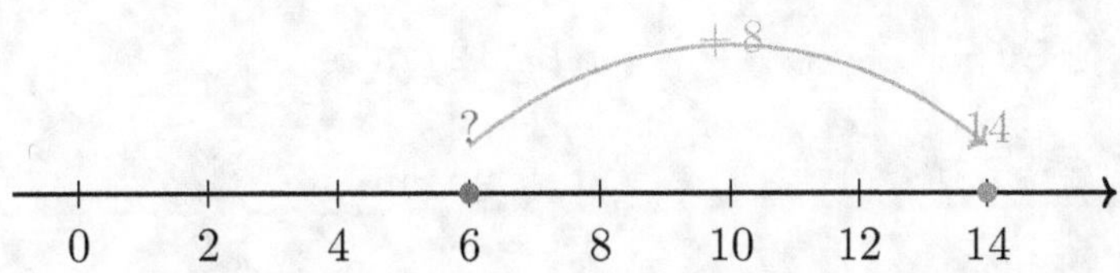

(A) $x + 8 = 14$, so $x = 6$

(B) $x - 8 = 14$, so $x = 22$

(C) $8x = 14$, so $x = 1.75$

(D) $x + 14 = 8$, so $x = -6$

17. Which of the following values is NOT a solution to $y \geq 3$?

(A) $y = 3$

(B) $y = 4$

(C) $y = 100$

(D) $y = 2.5$

18. Describe the graph of $n \leq 5$.

(A) Open circle at 5, shade left

(B) Closed circle at 5, shade left

(C) Open circle at 5, shade right

(D) Closed circle at 5, shade right

19. The number of inches i equals 12 times the number of feet f: $i = 12f$. How many inches are in 5 feet?

(A) 17 inches

(B) 48 inches

(C) 60 inches

(D) 125 inches

20. In a parallelogram, which measurement is used as the height?

(A) The slanted side

(B) The diagonal

(C) The perpendicular distance between the base and its opposite side

(D) The longest side

Find more at
ViewMath.com/GA-Grade6

ViewMath.com

21. A rectangular prism has a volume of 120 cm^3. Its length is 10 cm and width is 4 cm. What is the height?

(A) 3 cm

(B) 12 cm

(C) 6 cm

(D) 30 cm

22. A rectangle has vertices $(-5, 4)$, $(3, 4)$, $(3, -2)$, and $(-5, -2)$. What is the perimeter?

(A) 14 units

(B) 22 units

(C) 28 units

(D) 48 units

23. What is the area of the shaded right triangle on the coordinate plane?

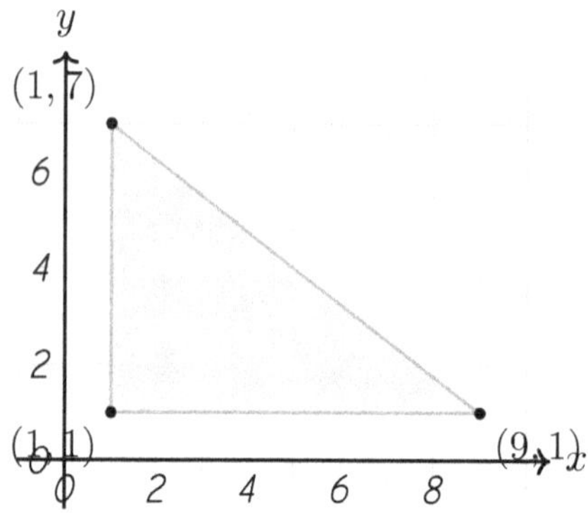

(A) 48 square units

(B) 24 square units

(C) 36 square units

(D) 12 square units

24. A cube has a surface area of 216 cm^2. What is the edge length?

Your Answer

25. Data set: $22, 24, 25, 25, 26, 27, 45$. Which value is most likely an outlier?

(A) 22

(B) 25

(C) 27

(D) 45

26. Data: $11, 13, 15, 17, 19, 21, 23$. What are the mean and median?

(A) Mean $= 17$, Median $= 17$

(B) Mean $= 17$, Median $= 15$

(C) Mean $= 15$, Median $= 17$

(D) Mean $= 19$, Median $= 17$

27. If every value in a data set is increased by 5, what happens to the range?

(A) It increases by 5.

(B) It decreases by 5.

(C) It stays the same.

(D) It doubles.

28. A box plot has a very long right whisker and a short left whisker. What does this suggest?

(A) The data is symmetric.

(B) The data is skewed right (has high outliers or a long tail to the right).

(C) The data is skewed left.

(D) All values are the same.

29. The table below shows statistics for two math classes on a unit test.

Statistic	Period 1	Period 3
Mean	76	82
Median	78	81
Range	40	18
IQR	16	8

Write a paragraph comparing the two classes. Use specific numbers to support your claims.

Your Answer:

Find more at
ViewMath.com/GA-Grade6

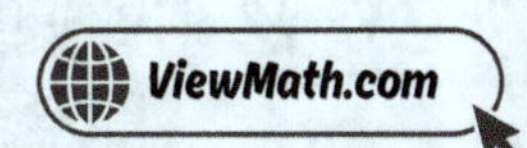

30. *A standard number cube is rolled. What is the probability of rolling a 5, expressed as a decimal rounded to the nearest hundredth?*

(A) 0.50

(B) 0.20

(C) 0.17

(D) 0.83

1. A class votes on two activities: hiking and swimming. The ratio of votes for hiking to swimming is 3 : 2. There are 25 votes total. How many voted for swimming?

Your Answer:

2. A car travels 240 miles using 8 gallons of gas. What is the unit rate in miles per gallon?

(A) 8

(B) 32

(C) 30

(D) 248

3. It takes 6 cups of flour to make 4 loaves of bread. How many cups of flour for 10 loaves?

(A) 12

(B) 10

(C) 15

(D) 20

Find more at
ViewMath.com/GA-Grade6

4. *The table and partial graph below show a ratio relationship.*

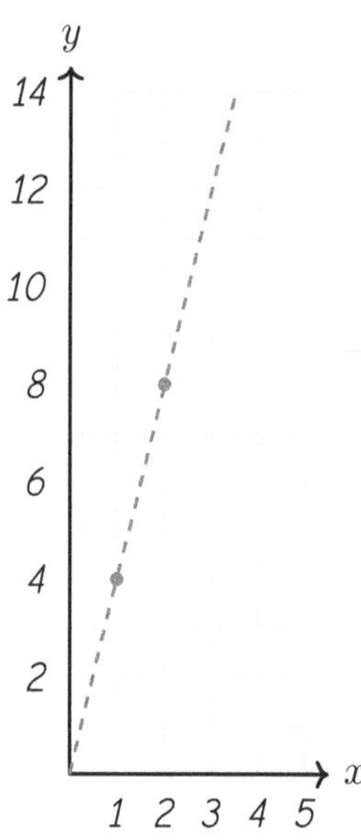

x	y
1	4
2	8
3	?

What is the missing value of y when $x = 3$?

(A) 10

(B) 12

(C) 14

(D) 16

5. *A map scale says 1 cm = 5 km. Two cities are 8 cm apart on the map. What is the real distance?*

(A) 13 km

(B) 5 km

(C) 40 km

(D) 80 km

6. *Convert 2 hours, 30 minutes to minutes, then to seconds.*

Your Answer:

7. What is $\dfrac{2}{3} \div \dfrac{4}{5}$?

   (A) $\dfrac{8}{15}$ (B) $\dfrac{6}{5}$

   (C) $\dfrac{5}{6}$ (D) $\dfrac{2}{15}$

8. In the standard long-division algorithm, the four repeating steps in order are: Divide, _________, Subtract, Bring Down. What is the missing step?

   (A) Estimate   (B) Multiply

   (C) Check   (D) Add

9. Which real-world situation is best described by a positive number?

   (A) A debt of \$50   (B) 20 feet below sea level

   (C) A gain of 6 yards in football   (D) A temperature of $10°$ below zero

10. A point on a number line is $\dfrac{3}{5}$ of the way from 0 to -5. What number does the point represent?

Your Answer:

11. Write an expression for: "9 fewer than a number z."

Your Answer:

12. In the expression $6(n + 2)$, what are the two factors?

   (A) 6 and n (B) 6 and 2

   (C) n and 2   (D) 6 and $(n + 2)$

Find more at
ViewMath.com/GA-Grade6

13. Evaluate $m^2 + 2m + 1$ when $m = 4$

Your Answer

14. Simplify: $4(3y - 2)$

(A) $12y - 2$ (B) $12y - 8$

(C) $7y - 2$ (D) $7y - 6$

15. A store sells notebooks for \$3 each and pens for \$1 each. Which expression gives the total cost for n notebooks and p pens?

(A) $3 + n + 1 + p$ (B) $3n + p$

(C) $4np$ (D) $n + 3p$

16. Solve: $\dfrac{w}{8} = 9$

Your Answer

17. Which phrase matches the inequality $t \leq 30$?

(A) The temperature is more than 30 degrees   (B) The temperature is exactly 30 degrees

(C) The temperature is no more than 30 degrees   (D) The temperature is at least 30 degrees

18. Which inequality matches a graph with a closed circle at 3 and shading to the right?

(A) $x > 3$ (B) $x < 3$

(C) $x \geq 3$ (D) $x \leq 3$

Find more at
ViewMath.com/GA-Grade6

ViewMath.com

19. In the equation $y = x + 4$, which statement is true?

(A) x is the dependent variable

(B) y is always 4

(C) x is the independent variable

(D) y decreases as x increases

20. A trapezoid has bases $b_1 = 4$ cm and $b_2 = 4$ cm with height 6 cm. What shape is this trapezoid actually equivalent to?

(A) A triangle

(B) A parallelogram

(C) A circle

(D) A pentagon

21. A box is 2.5 cm long, 4 cm wide, and 6 cm tall. What is the volume?

Your Answer:

22. A rectangle has vertices $(1, 2)$, $(7, 2)$, $(7, -4)$, and $(1, -4)$. What is the length of the longer side?

(A) 5 units

(B) 6 units

(C) 7 units

(D) 8 units

23. A rectangle on the coordinate plane has area 54 square units. Two of its vertices are at $(-3, 2)$ and $(6, 2)$. What is the width of the rectangle?

(A) 6 units

(B) 9 units

(C) 3 units

(D) 18 units

Find more at
ViewMath.com/GA-Grade6

24. *What is the surface area of a rectangular prism with length 6 cm, width 4 cm, and height 3 cm?*

(A) $72\ cm^2$ (B) $108\ cm^2$

(C) $54\ cm^2$ (D) $96\ cm^2$

25. *Data:* $5, 5, 6, 6, 6, 7, 7, 7, 7, 8, 8.$ *Where do the values cluster?*

(A) Around 5   (B) Around 6–7

(C) Around 8   (D) The data does not cluster.

26. *The mean of five numbers is 20. What is the sum of the five numbers?*

(A) 4   (B) 25

(C) 80   (D) 100

27. *Data:* $6, 8, 10, 12, 14.$ *The mean is 10. Find the MAD.*

Your Answer:

28. *A box plot has the median line closer to Q1 than to Q3. What does this mean?*

(A) The data is perfectly symmetric.   (B) The upper half of the data is more spread out than the lower half.

(C) The lower half is more spread out.   (D) The data has no outliers.

Find more at
ViewMath.com/GA-Grade6

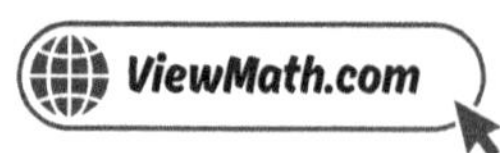

29. The two dot plots below show the daily tips (in dollars) earned by two servers at a restaurant.

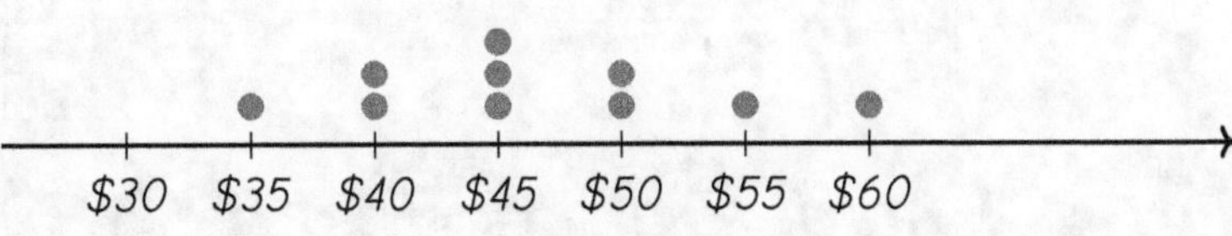

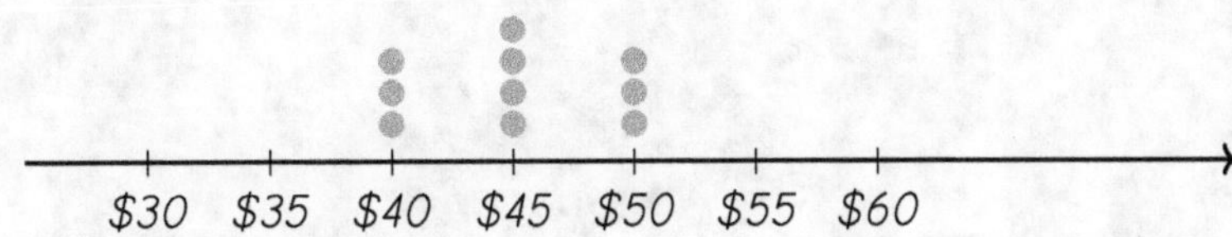

Find the median for each server. Which server earns more on a typical day? Which server's tips are more consistent?

Your Answer.

30. Event A has a probability of $\frac{2}{5}$ and Event B has a probability of $\frac{3}{4}$. Which statement is true?

(A) Event A is more likely than Event B   (B) Event B is more likely than Event A

(C) Both events are equally likely   (D) Neither event can happen

 # End of Practice Test 5 

Great job finishing the test!

 **My Score**

I got ___________ out of 30 questions right.

Check your answers in the **Answer Key** at the back of the book.

 Review any questions you missed. That's how we learn!

Check Your Score Online!

Visit **ViewMath Academy** to enter your answers and see which topics you need to review. You can also explore lessons, take quizzes, track your scores, and save your progress!

viewmath.com/score/6.1.GA.20

Or go to viewmath.com/score and enter code: 6.1.GA.20

6

Practice Test 6

 30 Questions

✏ Before You Start ✏

- ✔ **Read each question carefully** before choosing your answer.
- ✔ **Show your work** on scratch paper when you need to.
- ✔ **Skip hard questions** and come back to them later.
- ✔ **Check your answers** when you're done.
- ✔ **Take your time** — there's no rush!

⭐ You've Got This! ⭐

Do your best and show what you know!

1. A smoothie recipe uses 3 bananas **for every** 4 cups of yogurt. Write the ratio of bananas to yogurt. Then write the ratio of yogurt to bananas.

Your Answer

2. Brand X sells 5 notebooks for $8.75. Brand Y sells 3 notebooks for $4.50. Which is the better deal?

(A) Brand X at $1.75 each

(B) Brand Y at $1.50 each

(C) Brand X at $1.50 each

(D) Brand Y at $1.75 each

3. A teacher hands out 3 pencils for every 2 students. There are 16 students. How many pencils does she need?

(A) 18

(B) 24

(C) 20

(D) 32

4. A car travels at 30 miles per hour. Which set of points represents this ratio?

(A) $(1, 30)$, $(2, 60)$, $(3, 90)$

(B) $(30, 1)$, $(60, 2)$, $(90, 3)$

(C) $(1, 30)$, $(2, 50)$, $(3, 90)$

(D) $(1, 3)$, $(2, 6)$, $(3, 9)$

5. Which strategy is best for solving "how much for one?" problems?

(A) Tape diagram

(B) Unit rate

(C) Ratio table

(D) Equation

6. Convert 7,000 grams to kilograms.

(A) 0.7 kg

(B) 7 kg

(C) 70 kg

(D) 700 kg

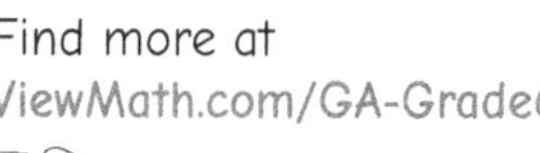

7. A recipe uses $\frac{2}{3}$ cup of sugar per batch. You have 4 cups of sugar. How many batches can you make?

(A) $\frac{8}{3}$

(B) 2

(C) 6

(D) 12

8. A student divides $3{,}612 \div 12$ and gets 31. The correct answer is 301. What mistake did the student most likely make?

(A) The student forgot to subtract

(B) The student skipped placing a zero in the quotient

(C) The student used the wrong divisor

(D) The student multiplied instead of dividing

9. The temperature outside is $-12°F$. What does this mean?

(A) 12 degrees above zero

(B) 12 degrees below zero

(C) Exactly zero degrees

(D) 12 degrees above freezing

10. Which list shows the numbers in order from least to greatest?

(A) $-0.5,\ -1.2,\ 0.3,\ 1.5$

(B) $-1.2,\ -0.5,\ 0.3,\ 1.5$

(C) $0.3,\ -0.5,\ 1.5,\ -1.2$

(D) $1.5,\ 0.3,\ -0.5,\ -1.2$

11. Which expression represents "6 more than half of a number n"?

(A) $6 + 2n$

(B) $\frac{n+6}{2}$

(C) $\frac{n}{2} + 6$

(D) $n + 3$

12. Look at the expression map below. Which label correctly identifies Part C?

$$\underbrace{5}_{A} \underbrace{x}_{B} + \underbrace{3}_{C}$$

(A) A variable

(B) A coefficient

(C) A constant

(D) A factor

13. Evaluate $\dfrac{x^2}{4}$ when $x = 8$.

Your Answer

14. Simplify: $3(x + 4) + 2x$

(A) $5x + 4$

(B) $5x + 12$

(C) $6x + 12$

(D) $3x + 6$

15. Emma earns \$8 per hour babysitting. She also got a \$15 tip. Which expression gives her total earnings for h hours?

(A) $8 + 15h$

(B) $23h$

(C) $8h + 15$

(D) $8h - 15$

16. Solve: $x + 3.5 = 10$

(A) $x = 6.5$

(B) $x = 13.5$

(C) $x = 7.5$

(D) $x = 3.5$

17. Write an inequality: A suitcase must weigh less than 50 pounds. Let w = weight.

Your Answer:

18. Pick two values that ARE solutions to $x > -1$ and one value that is NOT.

Your Answer:

19. A car wash charges $10 per car. Which variable goes on the x-axis when graphing?

(A) Total cost

(B) Number of cars

(C) Price per car

(D) Profit

20. A kite-shaped sign can be split into two triangles, each with base 6 in and height 8 in. What is the total area of the sign?

Your Answer:

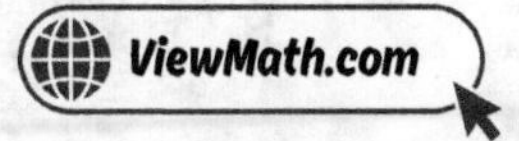

21. If you double the height of a rectangular prism but keep the length and width the same, what happens to the volume?

   (A) The volume stays the same.

   (B) The volume doubles.

   (C) The volume triples.

   (D) The volume quadruples.

22. Points $A(-1, -4)$ and $B(-1, 8)$ form a vertical segment. What is its length?

   Your Answer

23. A rectangle on the coordinate plane has an area of 56 square units. Its length is 8 units. What is the width?

   (A) 6 units

   (B) 7 units

   (C) 8 units

   (D) 48 units

24. How many faces does a rectangular prism have?

   (A) 4

   (B) 5

   (C) 6

   (D) 8

25. Data: $15, 16, 16, 17, 17, 17, 18, 100$. How does the outlier 100 affect the center?

   (A) It pulls the center toward 100, making it higher than typical values.

   (B) It has no effect on the center.

   (C) It makes the center equal to 100.

   (D) It pulls the center lower.

Find more at
ViewMath.com/GA-Grade6

26. Data: $8, 8, 9, 10, 10, 11, 50$. Which is the better measure of center?

(A) Mean, because it uses all the data.

(B) Median, because the outlier 50 pulls the mean too high.

(C) Neither is useful.

(D) Both are equally good.

27. Data: $1, 2, 3, 4, 5, 6, 100$. What is the range?

(A) 5

(B) 6

(C) 94

(D) 99

28. Data (in order): $3, 5, 7, 9, 11, 13, 15$. What is Q1?

(A) 3

(B) 5

(C) 7

(D) 9

29. When should you use the median and IQR instead of the mean and MAD?

(A) When the data is perfectly symmetric.

(B) When the data has outliers or is skewed.

(C) When every value is the same.

(D) When the data set is very small.

30. Which of the following best describes the range of all probabilities?

(A) From -1 to 1

(B) From 0 to 1

(C) From 0 to 100

(D) From 1 to 10

⭐ *End of Practice Test 6* ⭐

Great job finishing the test!

 *My Score*

I got _____________ out of 30 questions right.

Check your answers in the Answer Key at the back of the book.

💡 *Review any questions you missed. That's how we learn!*

📊 *Check Your Score Online!*

Visit **ViewMath Academy** *to enter your answers and see which topics you need to review. You can also explore lessons, take quizzes, track your scores, and save your progress!*

viewmath.com/score/6.1.GA.21

Or go to viewmath.com/score and enter code: 6.1.GA.21

Practice Test 7

 30 Questions

✏️ Before You Start ✏️

- ✔ **Read each question carefully** before choosing your answer.
- ✔ **Show your work** on scratch paper when you need to.
- ✔ **Skip hard questions** and come back to them later.
- ✔ **Check your answers** when you're done.
- ✔ **Take your time** — there's no rush!

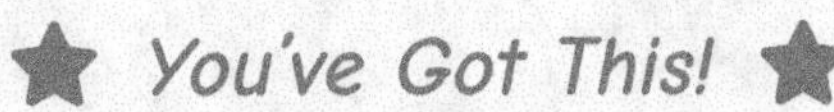

⭐ You've Got This! ⭐

Do your best and show what you know!

1. A lemonade stand sells 7 cups of lemonade **for each** 2 cups of iced tea. If they sold 21 cups of lemonade, how many cups of iced tea did they sell?

Your Answer:

2. A garden hose fills a 150-gallon tank in 5 hours. What is the unit rate?

(A) 25 gallons per hour

(B) 30 gallons per hour

(C) 35 gallons per hour

(D) 750 gallons per hour

3. Complete the ratio table for the ratio 5 : 8.

5	8
10	?
?	32

Your Answer:

4. A line passes through the origin and the point $(5, 15)$. What is the ratio x to y?

(A) 1 : 5

(B) 5 : 1

(C) 1 : 3

(D) 3 : 1

5. Two families split a \$350 dinner bill in a 3 : 4 ratio. How much does each family pay?

Your Answer:

Find more at
ViewMath.com/GA-Grade6

 # End of Practice Test 4 

Great job finishing the test!

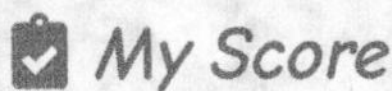 My Score

I got _____________ out of 30 questions right.

Check your answers in the **Answer Key** at the back of the book.

Review any questions you missed. That's how we learn!

Check Your Score Online!

Visit **ViewMath Academy** to enter your answers and see which topics you need to review. You can also explore lessons, take quizzes, track your scores, and save your progress!

viewmath.com/score/6.1.GA.19

Or go to viewmath.com/score and enter code: 6.1.GA.19

Practice Test 5

📋 30 Questions

✏️ Before You Start ✏️

- ✔ **Read each question carefully** before choosing your answer.
- ✔ **Show your work** on scratch paper when you need to.
- ✔ **Skip hard questions** and come back to them later.
- ✔ **Check your answers** when you're done.
- ✔ **Take your time** — there's no rush!

⭐ You've Got This! ⭐

Do your best and show what you know!

6. How many inches are in 5 feet?

   (A) 50   (B) 55

   (C) 60   (D) 72

7. A carpenter has $2\frac{1}{2}$ feet of wood. Each shelf needs $\frac{5}{8}$ of a foot. How many shelves can the carpenter cut?

   Your Answer:

8. What is $4{,}752 \div 12$?

   (A) 396   (B) 369

   (C) 406   (D) 39 R6

9. A submarine is 400 feet below sea level. Which integer represents the submarine's position?

   (A) 400   (B) −400

   (C) 0   (D) −40

10. Between which two consecutive integers is $\frac{11}{4}$ located on a number line?

    (A) 1 and 2   (B) 2 and 3

    (C) 3 and 4   (D) 0 and 1

11. Write a word phrase for the expression $5x + 3$.

    Your Answer:

Find more at
ViewMath.com/GA-Grade6

12. Which expression has exactly 4 terms?

(A) $2x + 3y$

(B) $a + b + c$

(C) $5m - 2n + 7 + m$

(D) $4p$

13. The perimeter of a rectangle is $P = 2l + 2w$. What is P when $l = 9$ and $w = 4$?

(A) 13

(B) 22

(C) 26

(D) 36

14. Simplify: $5p + 3 + 2p + 4p - 1$

(A) $11p + 2$

(B) $7p + 2$

(C) $11p + 4$

(D) $12p$

15. A taxi charges \$3 plus \$2 per mile. Write an expression for the cost of a ride that is m miles long.

Your Answer:

16. Solve: $5n = 45$

(A) $n = 9$

(B) $n = 40$

(C) $n = 50$

(D) $n = 225$

17. Which inequality represents "a number x is greater than 7"?

(A) $x < 7$

(B) $x > 7$

(C) $x \leq 7$

(D) $x = 7$

Find more at
ViewMath.com/GA-Grade6

ViewMath.com

18. A student graphs $x > 2$ with a closed circle at 2 and shading to the right. What is the error?

(A) Should shade to the left   (B) Should use an open circle at 2

(C) Should use a circle at 0 instead   (D) There is no error

19. A phone battery loses 5% each hour. Which is the independent variable?

(A) Battery percentage   (B) Phone model

(C) Number of hours   (D) Screen brightness

20. A trapezoid has an area of 84 cm^2 and bases of 10 cm and 14 cm. What is the height?

Your Answer:

21. A toy box is 3 ft long, 2 ft wide, and 2 ft tall. What is the volume?

(A) 7 ft^3 (B) 12 ft^3

(C) 6 ft^3 (D) 24 ft^3

22. To find the length of a horizontal side on the coordinate plane, you subtract the:

(A) y-coordinates and take the absolute value   (B) x-coordinates and take the absolute value

(C) x-coordinate from the y-coordinate   (D) coordinates and divide by 2

23. A rectangle has vertices $(-4, -3)$, $(6, -3)$, $(6, 2)$, and $(-4, 2)$. What is the area?

Your Answer:

Find more at
ViewMath.com/GA-Grade6

24. *Which net below could fold into a cube?*

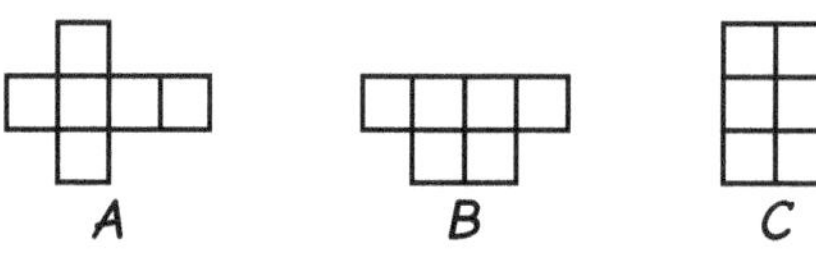

(A) Net A only

(B) Net B only

(C) Net C only

(D) Nets A and B

25. *Two students collected data. Student A: range = 5. Student B: range = 40. Whose data is more consistent?*

Your Answer

26. *The mean of 6 quiz scores is 15. If one score of 9 is removed, what is the mean of the remaining 5 scores?*

(A) 14.2

(B) 15

(C) 16.2

(D) 18

27. *The table below shows the hourly wages of 7 employees at a store.*

Employee	A	B	C	D	E	F	G
Wage ($)	9	10	10	12	14	15	30

Find the range and IQR. Explain which measure better describes the typical spread of wages.

Your Answer

28. *Which five values make up the five-number summary?*

(A) *Mean, median, mode, range, IQR*

(B) *Minimum, Q1, median, Q3, maximum*

(C) *Q1, Q2, Q3, Q4, Q5*

(D) *Mean, Q1, Q3, range, MAD*

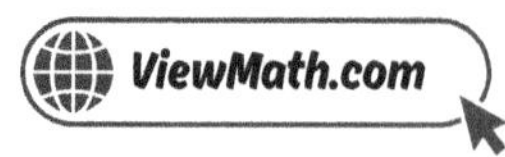

29. *Data:* 10, 12, 14, 16, 18, 20, 22. *The mean is 16. A student says the data is "very spread out." Is this correct?*

(A) *Yes, the range is 12.*

(B) *No, the range (12) is moderate and the values increase by a steady 2.*

(C) *Yes, because there are 7 values.*

(D) *No, because the mean equals the median.*

30. *A bag contains 8 marbles: 2 are yellow. What is the probability of drawing a yellow marble, written as a decimal?*

(A) 0.8

(B) 0.28

(C) 0.2

(D) 0.25

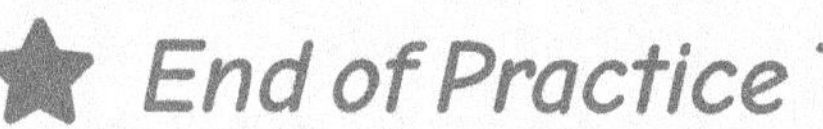  # ⭐ End of Practice Test 7 ⭐

Great job finishing the test!

☑ My Score

I got _____________ out of 30 questions right.

Check your answers in the **Answer Key** at the back of the book.

💡 Review any questions you missed. That's how we learn!

📊 Check Your Score Online!

Visit **ViewMath Academy** to enter your answers and see which topics you need to review. You can also explore lessons, take quizzes, track your scores, and save your progress!

viewmath.com/score/6.1.GA.22

Or go to viewmath.com/score and enter code: 6.1.GA.22

5. A printing company prints 250 flyers in 10 minutes. How long will it take to print 1,000 flyers?

> *Your Answer:*

6. The bar model below compares different length units.

1 yard		
1 ft	1 ft	1 ft

If a rope is 7 yards long, how many feet is it?

(A) 10

(B) 14

(C) 21

(D) 28

7. Kai says $\dfrac{2}{7} \div \dfrac{5}{6} = \dfrac{12}{35}$. Is Kai correct?

(A) No, the answer is $\dfrac{10}{42}$

(B) Yes, $\dfrac{12}{35}$ is correct

(C) No, the answer is $\dfrac{35}{12}$

(D) No, the answer is $\dfrac{7}{15}$

8. What is $7,344 \div 24$?

(A) 36

(B) 360

(C) 306

(D) 3,006

9. A scuba diver is 75 feet below sea level. Which integer represents the diver's depth?

(A) 75

(B) −75

(C) +75

(D) 0

10. *Order from least to greatest:* $-\dfrac{2}{3}, \; -0.75, \; \dfrac{1}{6}, \; -\dfrac{1}{2}.$

Your Answer

---

11. *The table below shows the relationship between a word phrase and an expression. Which expression belongs in the blank?*

Word Phrase	Expression
A number plus 4	$n + 4$
3 times a number	$3n$
8 less than a number	?
A number divided by 2	$n \div 2$

(A) $8 - n$

(B) $n - 8$

(C) $8n$

(D) $n + 8$

---

12. *What is the coefficient of d in the expression $15 - d + 3$?*

(A) 0

(B) 1

(C) -1

(D) 15

---

13. *A parking garage charges $5 plus $3 per hour. How much does it cost to park for 6 hours? Use the expression $5 + 3h$.*

Your Answer

Find more at
ViewMath.com/GA-Grade6

ViewMath.com

14. The rectangle below has the dimensions shown. Write a simplified expression for its perimeter.

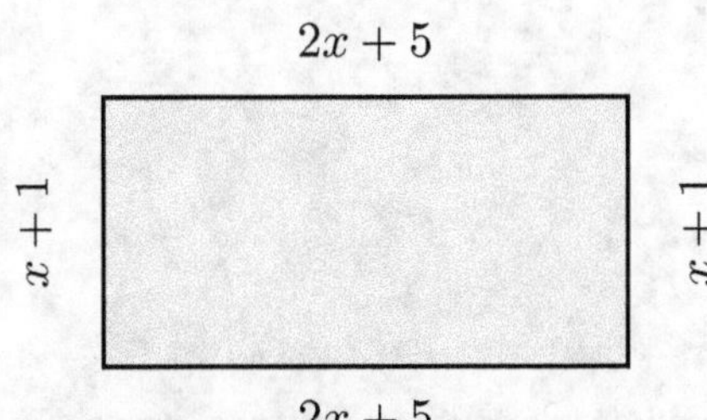

Your Answer:

15. Marcus has n baseball cards. He buys 12 more and then gives 5 to his friend. Write an expression for how many cards he has now.

Your Answer:

16. Solve: $15a = 105$

Your Answer:

17. Is $n = 4$ a solution to $n > 4$? Is $n = 4$ a solution to $n \geq 4$? Explain both.

Your Answer:

18. Which of these numbers is a solution to $x \geq -3$?

(A) -4

(B) -3.5

(C) -3

(D) -100

Find more at
ViewMath.com/GA-Grade6

ViewMath.com

19. A bathtub drains at 3 gallons per minute. It starts with 45 gallons. Write an equation for the water w remaining after t minutes. When is the tub empty?

Your Answer:

20. A student says the area of this parallelogram is 48 square units. The base is 8 and the slant side is 6. The actual height, shown by the dashed line, is 5. Is the student correct?

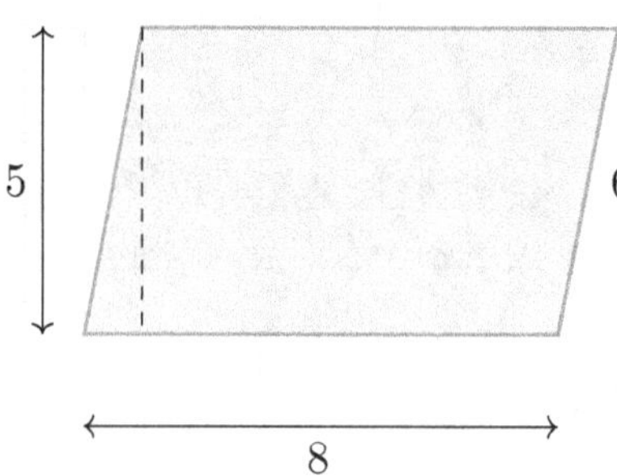

(A) Yes, because $8 \times 6 = 48$.

(B) No, the area is 40 because the height is 5, not 6.

(C) No, the area is 24 because you must divide by 2.

(D) Yes, because $\frac{1}{2} \times 8 \times 6 = 24$ and you double it.

21. A rectangular prism has length $\frac{1}{3}$ ft, width $\frac{1}{2}$ ft, and height 6 ft. What is the volume?

Your Answer:

22. What is the distance between $(-4, 3)$ and $(6, 3)$?

(A) 2 units

(B) 10 units

(C) 6 units

(D) 7 units

Find more at
ViewMath.com/GA-Grade6

ViewMath.com

23. A rectangle has vertices $(-3, 2)$, $(5, 2)$, $(5, -4)$, and $(-3, -4)$. What is the area?

(A) 24 square units   (B) 36 square units

(C) 48 square units   (D) 28 square units

24. A box is 15 in long, 6 in wide, and 4 in tall. How many square inches of cardboard are needed to make the box?

Your Answer:

25. Describe the shape of the data set: $1, 2, 3, 3, 4, 4, 4, 5, 5, 5, 5$.

Your Answer:

26. Data: $2, 3, 4, 5, 100$. The mean is 22.8. Why is the mean much higher than most of the values?

(A) There are only 5 values.   (B) The median is too low.

(C) The outlier 100 pulls the mean up.   (D) The values are all positive.

27. Data: $50, 55, 60, 65, 70$. The mean is 60. Find the MAD.

Your Answer:

28. Two box plots have the same median. Class A has IQR = 8 and Class B has IQR = 24. What can you conclude?

(A) Class A and Class B performed equally.   (B) Class A's scores are more spread out.

(C) Class B's middle 50% of scores are more   (D) The range of Class A is larger.
spread out.

29. *Class A: median = 90, IQR = 6. Class B: median = 82, IQR = 18. Compare the two classes.*

30. *A standard number cube is rolled. What is the probability of rolling a number less than or equal to 3?*

(A) $\dfrac{1}{3}$ (B) $\dfrac{3}{6}$

(C) $\dfrac{2}{6}$ (D) $\dfrac{4}{6}$

 # End of Practice Test 8 

Great job finishing the test!

✔️ My Score

I got ____________ out of 30 questions right.

*Check your answers in the **Answer Key** at the back of the book.*

💡 Review any questions you missed. That's how we learn!

📊 Check Your Score Online!

Visit **ViewMath Academy** to enter your answers and see which topics you need to review. You can also explore lessons, take quizzes, track your scores, and save your progress!

viewmath.com/score/6.1.GA.23

Or go to viewmath.com/score and enter code: 6.1.GA.23

Practice Test 9

✅ 30 Questions

✏️ Before You Start ✏️

- ✓ **Read each question carefully** before choosing your answer.
- ✓ **Show your work** on scratch paper when you need to.
- ✓ **Skip hard questions** and come back to them later.
- ✓ **Check your answers** when you're done.
- ✓ **Take your time** — there's no rush!

⭐ You've Got This! ⭐

1. The ratio of red to blue to green beads is $1 : 3 : 2$. There are 18 beads total. How many blue beads are there?

(A) 3

(B) 6

(C) 9

(D) 12

2. A printer prints 120 pages in 4 minutes. What is the unit rate?

(A) 4 pages per minute

(B) 30 pages per minute

(C) 120 pages per minute

(D) 480 pages per minute

3. Look at this ratio table. What is the missing value?

Apples	Oranges
3	5
6	?

(A) 8

(B) 10

(C) 15

(D) 11

4. Does the point $(3, 9)$ belong on the graph of the ratio $1 : 3$?

(A) No, because $3 + 9 \neq 1 + 3$.

(B) Yes, because $1 : 3 = 3 : 9$.

(C) No, because $3 \times 3 \neq 1$.

(D) Yes, because $3 + 9 = 12$.

5. A batch of trail mix uses 4 cups of granola for every 3 cups of raisins. If you have 12 cups of raisins, how many cups of granola do you need?

(A) 9

(B) 12

(C) 16

(D) 15

Find more at
ViewMath.com/GA-Grade6

ViewMath.com

6. A recipe needs 2.5 liters of water. How many milliliters is that?

(A) 25

(B) 250

(C) 2,500

(D) 25,000

7. Each serving of juice is $\frac{2}{5}$ cup. A jug holds $\frac{4}{5}$ cup. How many servings are in the jug?

(A) $\frac{8}{25}$

(B) $\frac{2}{5}$

(C) $\frac{5}{2}$

(D) 2

8. The long division below for $1{,}575 \div 5$ is partially completed. A digit in the quotient is missing.

$$
\begin{array}{r}
3\ \square\ 5 \\
5\ \overline{)\ 1\,5\,7\,5} \\
-1\,5 \\
\hline
7 \\
-5 \\
\hline
2\,5 \\
-2\,5 \\
\hline
0
\end{array}
$$

What digit belongs in the box? Explain how you know.

Your Answer

9. Which situation is best represented by the integer -15?

(A) A deposit of \$15 into a bank account

(B) A temperature of 15°F above zero

(C) An elevation of 15 feet above sea level

(D) A withdrawal of \$15 from a bank account

Find more at
ViewMath.com/GA-Grade6

ViewMath.com

10. *The number line below is divided into fourths between each pair of consecutive integers. Three points are labeled A, B, and C.*

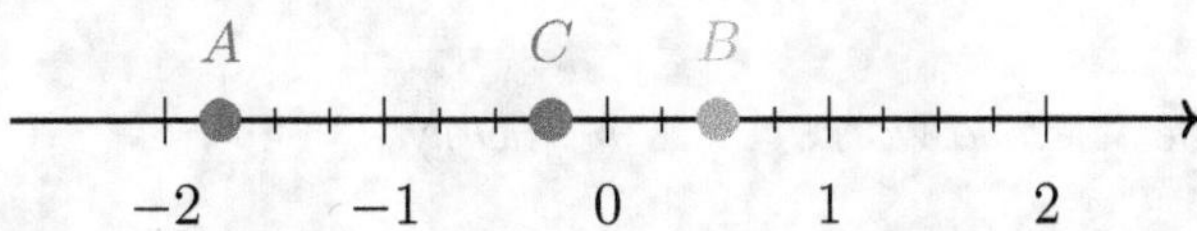

Part A: *Write the value of each point as a fraction or mixed number.*

Part B: *List the three values in order from least to greatest.*

Your Answer:

11. Which expression represents "the sum of 9 and the product of 4 and t"?

(A) $9 + 4 + t$

(B) $9 + 4t$

(C) $9(4 + t)$

(D) $9 \times 4t$

12. How many terms are in the expression $4x + 9 - 2y$?

(A) 2

(B) 3

(C) 4

(D) 5

13. Evaluate $6x - x^2$ when $x = 4$.

(A) 8

(B) 20

(C) -8

(D) 40

14. Simplify: $7n + 4n$

(A) $11n$

(B) $28n$

(C) $11n^2$

(D) $74n$

Find more at
ViewMath.com/GA-Grade6

ViewMath.com

15. In $P = 4s$, can s be any positive number, or is it one specific number?

   (A) Only $s = 1$

   (B) Only whole numbers

   (C) Any positive number — the formula works for all squares

   (D) Only $s = 4$

16. Solve: $\dfrac{t}{6} = 5$

   (A) $t = 11$

   (B) $t = 1$

   (C) $t = 30$

   (D) $t = 56$

17. Is $x = 5$ a solution to $x > 5$?

   (A) Yes, because 5 is equal to 5

   (B) Yes, because 5 is positive

   (C) No, because 5 is not greater than 5

   (D) No, because 5 is greater than 5

18. Which inequality matches a graph with an open circle at -2 and shading to the left?

   (A) $x > -2$

   (B) $x < -2$

   (C) $x \geq -2$

   (D) $x \leq -2$

19. In $c = 7p$, where c is total cost and p is the number of pizzas, what is the cost of 6 pizzas?

   (A) \$13

   (B) \$36

   (C) \$42

   (D) \$67

Find more at
ViewMath.com/GA-Grade6

20. A field is shaped like a parallelogram with base 25 m and height 14 m. Each bag of seed covers 50 m^2. How many bags are needed?

(A) 5 bags

(B) 6 bags

(C) 7 bags

(D) 8 bags

21. Two boxes have the same volume. Box A is $6 \times 4 \times 5$. Box B is $10 \times 3 \times h$. What is the height h of Box B?

(A) 2

(B) 4

(C) 6

(D) 8

22. What is the distance between the points $(2, 5)$ and $(2, -3)$?

(A) 2 units

(B) 5 units

(C) 8 units

(D) 3 units

23. A right triangle has vertices $(0, 0)$, $(10, 0)$, and $(10, 4)$. What is the area?

(A) 20 square units

(B) 40 square units

(C) 14 square units

(D) 7 square units

24. A cube has edges of 5 in. What is the surface area?

(A) 25 in^2

(B) 125 in^2

(C) 150 in^2

(D) 75 in^2

Find more at
ViewMath.com/GA-Grade6

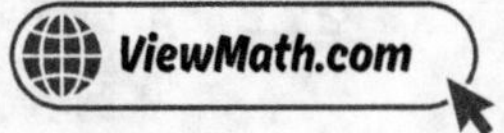

25. Look at the dot plot below.

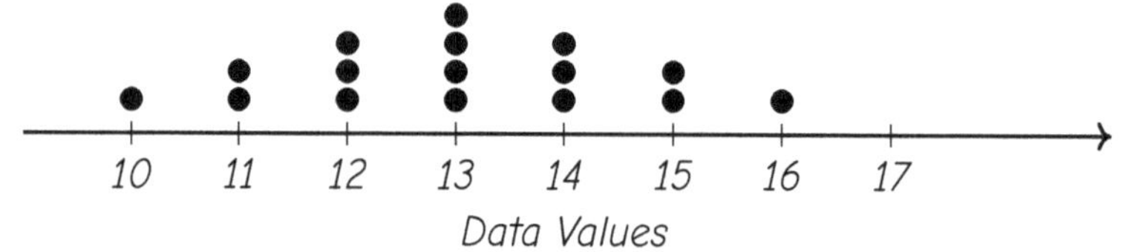

What is the shape of this data?

(A) Skewed to the right

(B) Skewed to the left

(C) Roughly symmetric

(D) Has a large gap in the middle

26. Four students' heights in inches are: $58, 60, 62, 64$. A fifth student who is 82 inches tall joins. What happens to the mean?

(A) It stays at 61.

(B) It increases to 65.2.

(C) It decreases.

(D) It increases to 82.

27. A number line below shows the positions of $Q1$, the median, and $Q3$ for a data set.

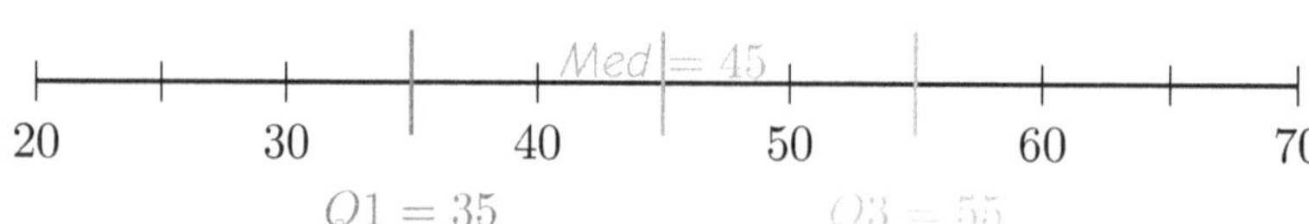

What is the IQR?

(A) 10

(B) 15

(C) 20

(D) 35

28. Data (in order): $2, 4, 6, 8, 10, 12$. What is the median?

(A) 6

(B) 7

(C) 8

(D) 10

Find more at
ViewMath.com/GA-Grade6

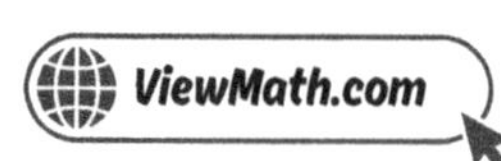

29. A report says: "The median score was 72 with an IQR of 15." What does this tell you?

(A) All scores are 72.

(B) The typical score is about 72, and the middle 50% of scores span 15 points.

(C) The highest score is 87.

(D) The lowest score is 57.

30. A box contains cards numbered 1 through 20. One card is drawn at random. What is the probability of drawing a number that is a multiple of 5? Write your answer as a fraction in simplest form.

Your Answer:

 # End of Practice Test 9 

Great job finishing the test!

 **My Score**

I got ___________ out of 30 questions right.

Check your answers in the **Answer Key** at the back of the book.

Review any questions you missed. That's how we learn!

Check Your Score Online!

Visit **ViewMath Academy** to enter your answers and see which topics you need to review. You can also explore lessons, take quizzes, track your scores, and save your progress!

viewmath.com/score/6.1.GA.24

Or go to viewmath.com/score and enter code: 6.1.GA.24

Practice Test 10

📋 30 Questions

✏️ Before You Start ✏️

- ✔ **Read each question carefully** before choosing your answer.
- ✔ **Show your work** on scratch paper when you need to.
- ✔ **Skip hard questions** and come back to them later.
- ✔ **Check your answers** when you're done.
- ✔ **Take your time** — there's no rush!

⭐ You've Got This! ⭐

Do your best and show what you know!

1. The tape diagram below shows the ratio of apple juice to orange juice in a drink.

Apple: ☐☐☐

Orange: ☐☐☐☐☐

If each part equals 4 ounces, how many total ounces of drink are there?

(A) 12

(B) 20

(C) 32

(D) 8

2. The table below shows the prices at two stores for packs of pencils.

	Number of Pencils	Price
Store A	10	$4.50
Store B	8	$3.20

Which store has the lower unit price per pencil?

(A) Store A at $0.45 per pencil

(B) Store B at $0.40 per pencil

(C) Store A at $0.40 per pencil

(D) They have the same unit price.

3. Which pair of ratios are equivalent?

(A) 2 : 3 and 4 : 9

(B) 2 : 3 and 6 : 9

(C) 2 : 3 and 8 : 9

(D) 2 : 3 and 3 : 2

Find more at
ViewMath.com/GA-Grade6

4. Which table matches a graph that passes through $(0,0)$, $(2,3)$, and $(4,6)$?

(A)

x	y
2	3
6	9

(B)

x	y
2	3
6	8

(C)

x	y
3	2
6	9

(D)

x	y
2	4
6	12

5. A fence uses posts in a ratio of 1 post for every 6 feet. How many posts are needed for a 42-foot fence?

(A) 6

(B) 7

(C) 8

(D) 36

6. The number line below shows a distance in centimeters.

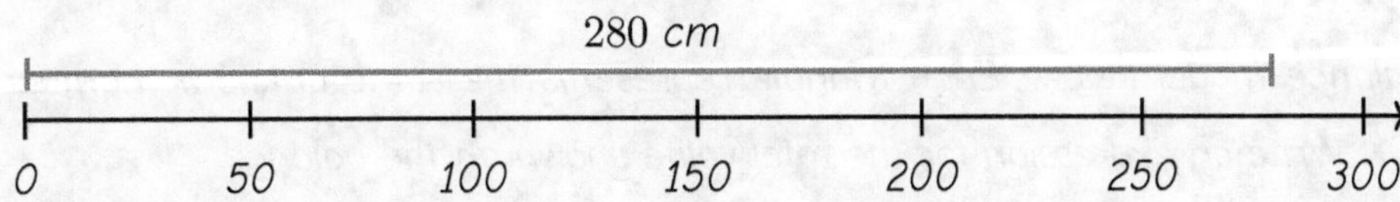

Part A: Convert 280 centimeters to meters.

Part B: Convert 280 centimeters to millimeters.

Part C: A doorway is 2 meters tall. Is 280 cm longer or shorter? By how much?

Your Answer:

7. Evaluate: $\dfrac{7}{10} \div \dfrac{2}{5}$

Your Answer:

8. *Compute* $6{,}048 \div 21$. *Then check your answer using multiplication.*

9. On a number line, where are negative numbers located?

A To the right of zero   B To the left of zero

C Above zero   D On top of zero

10. Between which two consecutive integers is $-\dfrac{5}{3}$ located on a number line?

A -1 and 0 B -2 and -1

C -3 and -2 D 0 and 1

11. Look at the balance model below. Each triangle represents the same unknown number x, and each small square represents 1. Write an expression for the total value shown on the balance.

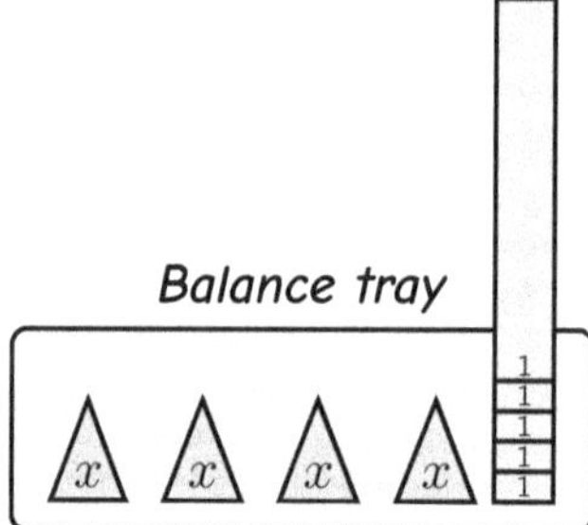

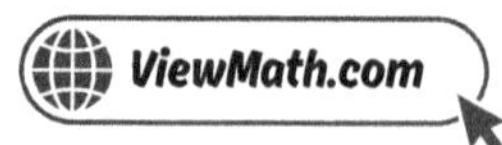

12. In the term $\dfrac{n}{5}$, what are the factors?

(A) n and 5 (B) n and $\dfrac{1}{5}$

(C) 1 and n (D) 5 and $\dfrac{1}{n}$

13. The table below shows the cost of renting a bike using the expression $C = 8 + 3h$, where h is the number of hours. Fill in the missing values A and B.

Hours (h)	Cost (C)
1	$11
2	A
3	$17
5	B

Your Answer:

14. Which expression is equivalent to $7(k - 3) + 5$?

(A) $7k - 16$ (B) $7k + 2$

(C) $7k - 26$ (D) $7k - 21$

15. You have 100 pages to read and read r pages each day. Which expression tells how many pages are left after 4 days?

(A) $100 + 4r$ (B) $100 - 4r$

(C) $4r - 100$ (D) $100r - 4$

Find more at
ViewMath.com/GA-Grade6

ViewMath.com

16. *Solve:* $m + 17 = 30$

Your Answer:

17. Which inequality represents "you need more than \$25 to buy the video game"?

(A) $d \leq 25$ (B) $d < 25$

(C) $d > 25$ (D) $d \geq 25$

18. Look at the number line below and write the inequality it represents.

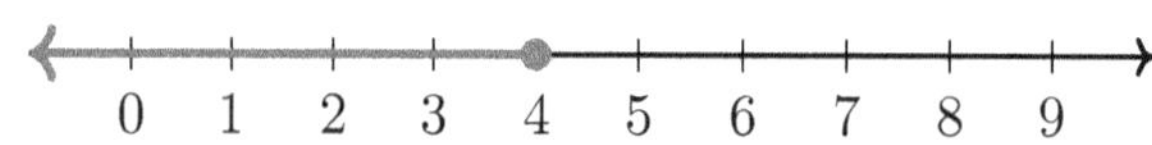

Your Answer:

19. Which equation represents this relationship: "The number of legs L is 4 times the number of dogs d"?

(A) $d = 4L$ (B) $L = d + 4$

(C) $L = 4d$ (D) $L = d \div 4$

20. A trapezoid has bases $b_1 = 5$ ft and $b_2 = 13$ ft, and height $h = 8$ ft. What is the area?

(A) $144\ ft^2$ (B) $72\ ft^2$

(C) $52\ ft^2$ (D) $36\ ft^2$

21. What is the volume of a rectangular prism with length 5 cm, width 3 cm, and height 4 cm?

(A) $12\ cm^3$ (B) $60\ cm^3$

(C) $30\ cm^3$ (D) $94\ cm^3$

Get Online

Find more at
ViewMath.com/GA-Grade6

ViewMath.com

22. A triangle on the coordinate plane has vertices $(0,0)$, $(8,0)$, and $(4,6)$. What is the length of the base along the x-axis?

(A) 4 units

(B) 6 units

(C) 8 units

(D) 10 units

23. A right triangle has vertices $(-3,-2)$, $(5,-2)$, and $(-3,4)$. What is the area?

Your Answer:

24. The diagram shows a rectangular prism with its dimensions. What is the area of the shaded face?

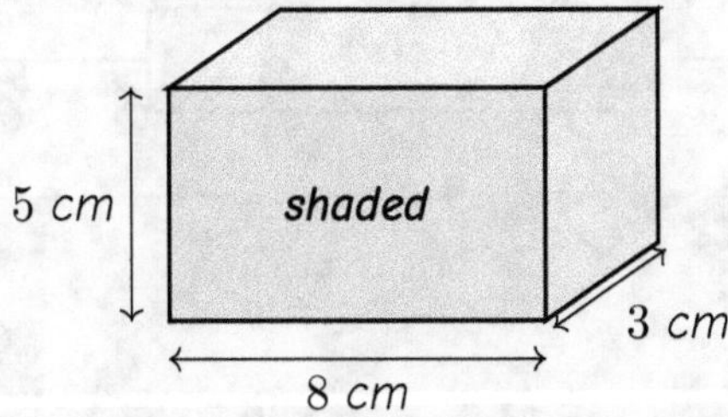

(A) $15\ cm^2$

(B) $24\ cm^2$

(C) $40\ cm^2$

(D) $120\ cm^2$

25. Student test times (minutes): $8, 9, 10, 10, 11, 12, 25$. A classmate says the typical time is about 12 minutes. Is this a good estimate?

(A) Yes, 12 is in the data set.

(B) Yes, because the outlier should be included.

(C) No, most times cluster around 9–11, so a typical value is closer to 10.

(D) No, the typical value is 25.

Find more at
ViewMath.com/GA-Grade6

26. A student scored 85, 90, 78, 92, and 95 on five tests. What score does the student need on a sixth test to have a mean of 90?

Your Answer

27. Data (in order): $4, 6, 8, 10, 12, 14, 16$. What is Q1?

(A) 4

(B) 6

(C) 8

(D) 10

28. Look at the box plot below.

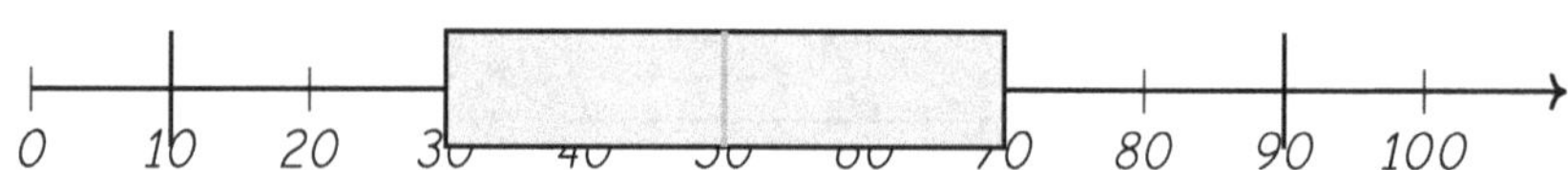

What is the IQR of this data?

(A) 20

(B) 40

(C) 50

(D) 80

29. When two data sets have very little overlap, what can you say?

(A) The groups are very similar.

(B) The groups are clearly different from each other.

(C) Both groups have the same center.

(D) Both groups have the same spread.

30. A bag holds 6 blue tokens, 4 red tokens, and 2 white tokens. What is the probability of drawing a blue token?

(A) $\dfrac{6}{10}$

(B) $\dfrac{1}{6}$

(C) $\dfrac{6}{12}$

(D) $\dfrac{4}{12}$

 *End of Practice Test 10* 

Great job finishing the test!

☑ My Score

I got _____________ out of 30 questions right.

*Check your answers in the **Answer Key** at the back of the book.*

💡 *Review any questions you missed. That's how we learn!*

📊 Check Your Score Online!

Visit **ViewMath Academy** to enter your answers and see which topics you need to review. You can also explore lessons, take quizzes, track your scores, and save your progress!

viewmath.com/score/6.1.GA.25

Or go to viewmath.com/score and enter code: 6.1.GA.25

Answer Key & Explanations

8

Practice Test 8

30 Questions

✏️ Before You Start ✏️

- ✔ **Read each question carefully** before choosing your answer.
- ✔ **Show your work** on scratch paper when you need to.
- ✔ **Skip hard questions** and come back to them later.
- ✔ **Check your answers** when you're done.
- ✔ **Take your time** — there's no rush!

⭐ You've Got This! ⭐

Do your best and show what you know!

1. The ratio of cats to dogs at a pet store is $3 : 4$. Each part in the tape diagram represents 2 animals. How many dogs are there?

(A) 6

(B) 8

(C) 3

(D) 4

2. The graph shows the number of laps a swimmer completes over time.

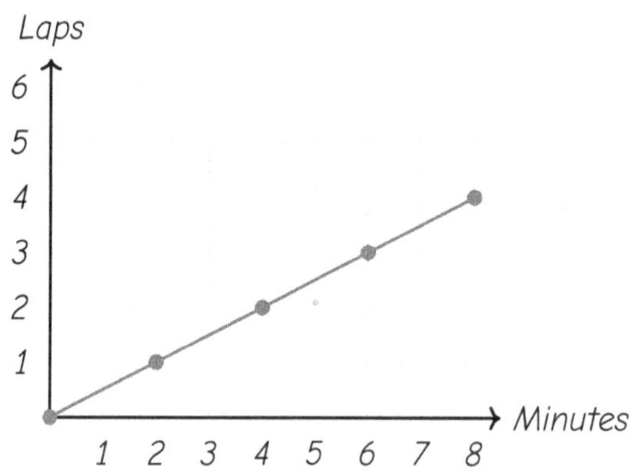

Part A: What is the swimmer's unit rate in laps per minute?

Part B: How many laps will the swimmer complete in 14 minutes?

Your Answer

3. A smoothie recipe uses 2 cups of strawberries for every 3 cups of yogurt. How many cups of yogurt are needed for 10 cups of strawberries?

Your Answer

4. A ratio graph passes through $(0, 0)$ and $(4, 7)$. What is y when $x = 8$?

(A) 11

(B) 14

(C) 15

(D) 12

Find more at
ViewMath.com/GA-Grade6

Answer Key

First try each test on your own, then check your work here.

☑ Practice Test 1 — Answer Key

1 5   **2** C   **3** C   **4** C   **5** C   **6** 8 full bottles   **7** A   **8** D

9 Neither positive nor negative   **10** $-1.5, \ -\frac{3}{4}, \ 0.2, \ \frac{1}{2}, \ \frac{5}{4}$ **11** B   **12** $A = 4, B = (x + 3)$

13 22   **14** B   **15** Answers vary. Example: You buy n notebooks at \$5 each and pay \$3 for shipping.

16 C   **17** Answers vary. Example: A classroom has at most 15 computers.

18 $x = 3$ IS a solution to $x \le 3$ because $3 \le 3$ is true. $x = 3$ is NOT a solution to $x < 3$ because $3 < 3$ is false.

19 B   **20** A   **21** C   **22** D   **23** B   **24** D   **25** A   **26** A   **27** B

28 Range $= 55$, IQR $= 30$ **29** A   **30** 0.625

💡 Time to Learn! 💡

Review the explanations below, **especially for the questions you missed**.

Understanding why each answer is correct builds stronger problem-solving skills.

Tip: Circle any questions you got wrong, then read their explanation carefully.

📖 Practice Test 1 — Detailed Explanations

1. Each part $= 20 \div 4 = 5$. Vegetables $= 1 \times 5 = 5$.

2. From the graph, the truck travels 50 miles in 2 hours. Unit rate: $50 \div 2 = 25$ miles per hour.

3. From the double number line, 9 scoops aligns with 15 cups of water.

4. $20 \div 5 = 4$. The unit rate is 4 (units of y per 1 unit of x).

5. Scale: 2 cm $= 15$ km, so 1 cm $= 7.5$ km. For 5 cm: $5 \times 7.5 = 37.5$ km.

6. 6 L $= 6{,}000$ mL. $6{,}000 \div 750 = 8$.

7. The number line shows 5 jumps of $\dfrac{1}{6}$ to reach $\dfrac{5}{6}$. This models $\dfrac{5}{6} \div \dfrac{1}{6} = 5$.

8. $1{,}344 \div 8 = 168$ pages per week. Check: $168 \times 8 = 1{,}344$.

9. Zero is the boundary between positive and negative. It is neither positive nor negative.

10. Convert to decimals: $\dfrac{1}{2} = 0.5$, $-\dfrac{3}{4} = -0.75$, 0.2, -1.5, $\dfrac{5}{4} = 1.25$. From least to greatest: $-1.5 < -0.75 < 0.2 < 0.5 < 1.25$.

11. Sandwiches cost $6s$ and drinks cost $2d$. Total: $6s + 2d$.

12. $4(x + 3) = 4 \times (x + 3)$. The two factors are 4 and $(x + 3)$.

Find more at
ViewMath.com/GA-Grade6

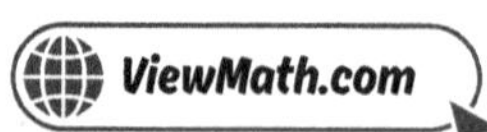

13   $3(8 + 2) - 8 = 3(10) - 8 = 30 - 8 = 22.$

14   Distribute: $3 \times x + 3 \times 5 = 3x + 15.$

15   $5n$ represents a cost per item and 3 represents a fixed fee.

16   Multiply both sides by 4: $m = 7 \times 4 = 28.$

17   Any situation where a quantity is 15 or fewer works.

18   $\leq$ includes the boundary value; $<$ does not.

19   In B, y is always 7 no matter what x is. The value of y does not change with x.

20   Rectangle: $8 \times 5 = 40.$ Triangle: $\frac{1}{2} \times 8 \times 3 = 12.$ Total: $40 + 12 = 52 \ cm^2.$

21   Volume measures 3-dimensional space, so it uses cubic units like cm^3, in^3, or ft^3.

22   $|4 - (-5)| = |9| = 9$ units.

23   Base $= |8 - 2| = 6.$ Height $= |7 - 1| = 6.$ Area $= \frac{1}{2} \times 6 \times 6 = 18$ square units.

24   $SA = 6 \times 3^2 = 6 \times 9 = 54 \ cm^2.$

25   Right-skewed data has most values clustered on the lower end with a few high values stretching to the right.

26   Total students $= 15.$ Median is the 8th value. In order: three 1's (3 so far), five 2's (8 so far). The 8th value is 2.

Find more at
ViewMath.com/GA-Grade6

27 $Q1$: median of $3, 5, 7 = 5$. $Q3$: median of $10, 12, 14 = 12$. $IQR = 12 - 5 = 7$.

28 Range $= 80 - 25 = 55$. $IQR = 65 - 35 = 30$.

29 Class A's MAD is 3, meaning scores are typically within 3 of the mean. Class B's scores typically deviate 9 points. Class A's student is more likely to be near 75.

30 $P(blue) = \dfrac{3}{8}$. $P(not\ blue) = 1 - \dfrac{3}{8} = \dfrac{5}{8} = 0.625$.

☑ Practice Test 2 — Answer Key

1 C

2 Part A: Lily earns \$12 per hour, Noah earns \$13 per hour.   Part B: Noah earns \$1 more per hour.

3 Part A: $2 : 1$;   Part B: 4 cups of sugar   **4** C   **5** B   **6** C   **7** B   **8** 301   **9** C

10 $\dfrac{1}{2}$ **11** C   **12** 5 and $(y + 8)$ **13** 30 square cm   **14** $6a + 10$ **15** A   **16** B

17 C   **18** C   **19** Independent: h (hours). Dependent: d (distance).   **20** B   **21** B

22 26 units   **23** B   **24** 5 ft   **25** D   **26** Mean ≈ 66, Median $= 66$ **27** C   **28** D

29 B   **30** C

💡 Time to Learn! 💡

Review the explanations below, especially for the questions you missed.

Understanding why each answer is correct builds stronger problem-solving skills.

Tip: Circle any questions you got wrong, then read their explanation carefully.

Find more at
ViewMath.com/GA-Grade6

📖 Practice Test 2 — Detailed Explanations

1 Girls have 6 parts $= 24$, so each part $= 24 \div 6 = 4$. Boys have 4 parts $= 4 \times 4 = 16$.

2 Lily: $\$60 \div 5 = \$12/hr$. Noah: $\$52 \div 4 = \$13/hr$. Noah earns $\$13 - \$12 = \$1$ more per hour.

3 Part A: From the graph, $(2, 1)$ shows 2 cups of flour for every 1 cup of sugar. Part B: Following the pattern, $(8, 4)$, so 4 cups of sugar.

4 At $(2, 5)$: $x : y = 2 : 5$. Check: $(4, 10)$ simplifies to $2 : 5$ as well.

5 $360 \div 12 = 30$ minutes.

6 $1 \; km = 1{,}000 \; m = 100{,}000 \; cm$. $10 \times 100{,}000 = 1{,}000{,}000 \; cm$.

7 $\dfrac{3}{5} \times \dfrac{5}{1} = \dfrac{15}{5} = 3$.

8 $45 \div 15 = 3$; bring down $1 \to 1 \div 15 = 0$ R1; bring down $5 \to 15 \div 15 = 1$. Answer: 301. The zero in the tens place is important. Check: $301 \times 15 = 4{,}515$.

9 Zero in a bank account means no savings and no debt. It is the boundary between having money (positive) and owing money (negative).

10 Halfway $= \dfrac{\frac{1}{4} + \frac{3}{4}}{2} = \dfrac{\frac{4}{4}}{2} = \dfrac{1}{2}$.

11 Equally distributing c cupcakes into 6 boxes means dividing: $c \div 6$.

12 $5(y + 8) = 5 \times (y + 8)$. The factors are 5 and $(y + 8)$.

Find more at
ViewMath.com/GA-Grade6

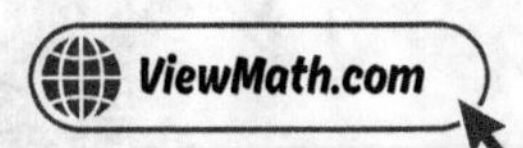

13. $A = \frac{1}{2} \times 10 \times 6 = \frac{1}{2} \times 60 = 30$ *square cm.*

14. $3a + 5a - 2a = 6a$ *and* $7 + 3 = 10.$ *Result:* $6a + 10.$

15. *Tickets:* $9p.$ *One bucket of popcorn:* $6.$ *Total:* $9p + 6.$

16. *Subtract* $15:$ $k = 32 - 15 = 17.$

17. *"Fewer than 20" means less than 20:* $y < 20.$

18. *Open circle at 0 means 0 is not included. Shading right means greater than:* $x > 0.$

19. *You choose the hours (independent). Distance depends on hours:* $d = 6h.$

20. *Area uses the height, not the slant side.* $A = 12 \times 6 = 72 \ m^2.$

21. *Total volume* $= 5 \times 3 \times 2 = 30 \ m^3.$ *Half-full:* $30 \div 2 = 15 \ m^3.$

22. *Length* $= |2 - (-6)| = 8.$ *Width* $= |2 - (-3)| = 5.$ *Perimeter* $= 2(8) + 2(5) = 26 \ units.$

23. *Decompose irregular shapes into simpler shapes (rectangles and triangles) whose areas you can compute, then add them.*

24. $2(2)(3) + 2(2)(h) + 2(3)(h) = 62.$ $12 + 4h + 6h = 62.$ $10h = 50.$ $h = 5 \ ft.$

25. *Most values are between 2 and 7, clustered around 4. The value 20 is far away from the rest, making it an outlier.*

Find more at
ViewMath.com/GA-Grade6

26. Values: $60, 64, 64, 66, 68, 68, 72$. Sum $= 462$. Mean $= 462 \div 7 = 66$. Median (4th of 7) $= 66$.

27. Range $= 25 - 8 = 17$.

28. Different data sets can have the same five-number summary. The five-number summary only captures five key values, not every data point.

29. Class X ranges from 25 to 40 (range $= 15$). Class Y ranges from 15 to 50 (range $= 35$). Class Y is much more spread out.

30. Event Y is positioned closest to 0.75 on the probability scale.

✔ Practice Test 3 — Answer Key

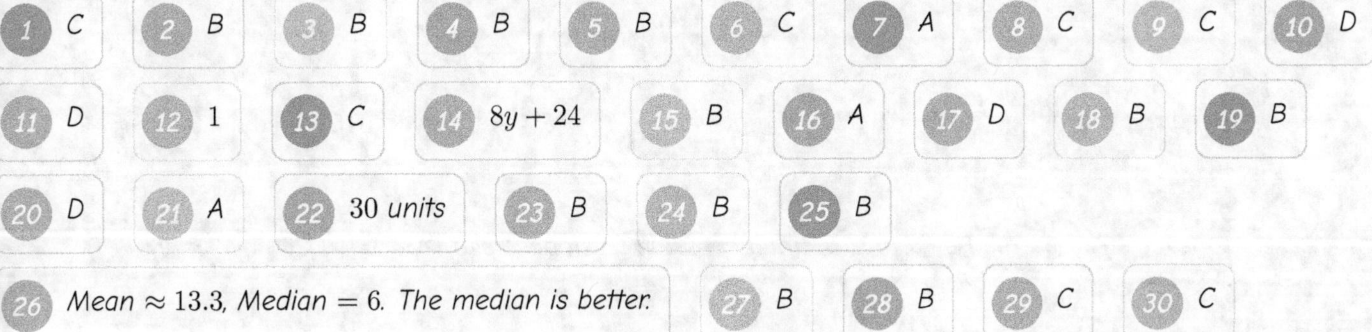

💡 Time to Learn! 💡

Review the explanations below, **especially for the questions you missed**.

Understanding why each answer is correct builds stronger problem-solving skills.

Tip: Circle any questions you got wrong, then read their explanation carefully.

Find more at
ViewMath.com/GA-Grade6

📖 Practice Test 3 — Detailed Explanations

1 The question asks flour to eggs. Flour $= 3$, eggs $= 2$. The ratio is $3 : 2$.

2 $\$9 \div 12 = \$0.75$ per muffin.

3 $7 \times 3 = 21$ bananas, so $\$2 \times 3 = \$6$.

4 The ratio is $6 : 2 = 3 : 1$. When $x = 15$: $y = 15 \div 3 = 5$.

5 Unit rate: $\$24 \div 3 = \$8$ per shirt. 10 shirts: $10 \times \$8 = \$80$.

6 $3 \times 5{,}280 = 15{,}840$ feet.

7 $\dfrac{1}{3} \times \dfrac{3}{2} = \dfrac{3}{6} = \dfrac{1}{2}$.

8 $15 \div 5 = 3$; bring down $7 \to 7 \div 5 = 1$ R2; bring down $5 \to 25 \div 5 = 5$. Answer: 315. Check: $315 \times 5 = 1{,}575$.

9 Above sea level is represented by a positive integer. 1,200 feet above sea level is $+1{,}200$ or simply 1,200.

10 Halfway between -1 and 0 is $\dfrac{-1 + 0}{2} = -\dfrac{1}{2}$. The number $-\dfrac{1}{2} = -0.5$ is at the midpoint.

11 Giving away 6 means subtracting: $x - 6$.

12 $4(a + b)$ is a single term — it is one product. (If you distribute it to $4a + 4b$, it would be 2 terms.)

13 $3^2 + 3 = 9 + 3 = 12$.

14. $8 \times y + 8 \times 3 = 8y + 24$.

15. $d = 60t$ means distance equals 60 miles per hour times the number of hours.

16. Subtract 24: $x = 50 - 24 = 26$.

17. "At most" means the value can equal the number or be less, so it should be $\leq$, not $<$.

18. "More than 75" is $t > 75$: open circle (not including 75), shade right (greater values).

19. 2 cups per batch, b batches: $f = 2b$.

20. $A = \frac{1}{2}(40 + 60)(30) = \frac{1}{2}(100)(30) = 1{,}500 \ m^2$.

21. $V = \frac{3}{4} \times 2 \times 4 = \frac{3}{4} \times 8 = 6 \ ft^3$.

22. Length $= |4 - (-5)| = 9$. Width $= |4 - (-2)| = 6$. Perimeter $= 2(9) + 2(6) = 30$ units.

23. Base $= |5 - 1| = 4$. Height $= |9 - 1| = 8$. Area $= \frac{1}{2} \times 4 \times 8 = 16$ square units.

24. One block: $2(3)(2) + 2(3)(4) + 2(2)(4) = 12 + 24 + 16 = 52 \ ft^2$. Two blocks: $52 \times 2 = 104 \ ft^2$.

25. Data set A range: $24 - 20 = 4$. Data set B range: $34 - 10 = 24$. Data set B is much more spread out.

26. Mean: $(3 + 4 + 5 + 6 + 7 + 8 + 60) \div 7 = 93 \div 7 \approx 13.3$. Median: 6. The outlier 60 inflates the mean. The median (6) is more typical.

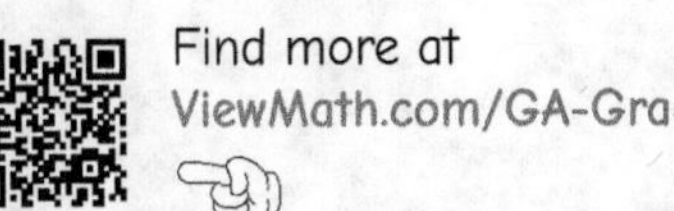
Find more at
ViewMath.com/GA-Grade6

27 MAD (Mean Absolute Deviation) is calculated by finding the average of the distances of each data value from the mean.

28 $IQR = Q3 - Q1 = 40 - 20 = 20.$

29 Same medians, but Class A's IQR (6) is much smaller than Class B's (18). Class A's scores are more tightly clustered.

30 Total balls: $5 + 3 + 2 = 10.$ $P(yellow) = \dfrac{2}{10}.$ $P(not\ yellow) = 1 - \dfrac{2}{10} = \dfrac{8}{10}.$

☑ Practice Test 4 — Answer Key

1 B   2 3.5 hours   3 C   4 Line A. It has a rate of 6 per 1, which is higher than 4 per 1.

5 B   6 D   7 A   8 168 cases   9 C   10 C   11 A   12 7 and 4   13 B

14 $11m + 2$ 15 B   16 A   17 D   18 B   19 C   20 C   21 A   22 C

23 B   24 6 cm   25 D   26 A   27 C   28 B

29 Period 3 performed better overall (mean 82 vs. 76, median 81 vs. 78) and was more consistent (IQR 8 vs. 16, rang

30 C

💡 Time to Learn! 💡

Review the explanations below, **especially for the questions you missed**.

Understanding why each answer is correct builds stronger problem-solving skills.

Tip: Circle any questions you got wrong, then read their explanation carefully.

Find more at
ViewMath.com/GA-Grade6

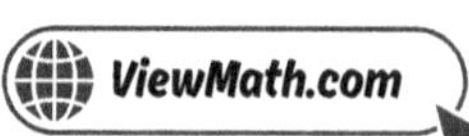

Practice Test 4 — Detailed Explanations

1 "3 pencils for every 1 eraser" means pencils to erasers is $3 : 1$.

2 $49 \div 14 = 3.5$ hours.

3 The ratio is $4 : 10 = 2 : 5$. Row 3 should be $12 : 30$ (since $4 \times 3 = 12$ and $10 \times 3 = 30$), but it shows $12 : 25$.

4 Line A goes up 6 for every 1 unit right; Line B goes up 4. Higher rise per unit means a faster rate.

5 Unit rate: $180 \div 3 = 60$ mph. In 7 hours: $60 \times 7 = 420$ miles.

6 $3 \times 4 = 12$ cups.

7 Dividing by $\frac{c}{d}$ is the same as multiplying by its reciprocal, $\frac{d}{c}$.

8 $5{,}376 \div 32 = 168$. Check: $168 \times 32 = 5{,}376$.

9 A negative number is less than zero. -4 is the only number less than zero in the list. Zero is neither positive nor negative.

10 Since $-2 < -1.5 < -1$, the number -1.5 lies between -2 and -1. It is exactly halfway between them.

11 $\frac{n}{3}$ is "n divided by 3." Adding 7 gives "7 more than n divided by 3."

12 7 and 4 are the terms without variables, so they are constants.

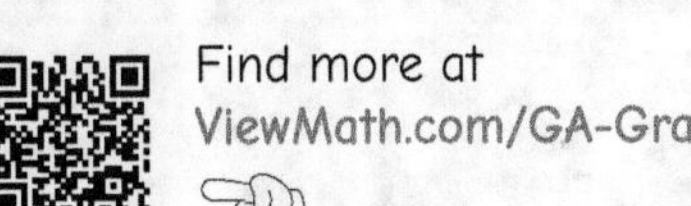

Find more at
ViewMath.com/GA-Grade6

13. $P = 2(7) + 2(3) = 14 + 6 = 20$ cm.

14. Distribute: $8m + 2 + 3m$. Combine: $8m + 3m = 11m$. Result: $11m + 2$.

15. A square has 4 equal sides, so its perimeter is 4 times the side length: $P = 4s$.

16. Start at unknown x, add 8, arrive at 14: $x + 8 = 14$, so $x = 6$.

17. $2.5 \geq 3$ is false. $2.5 < 3$, so it is not a solution.

18. $\leq$ means 5 is included (closed circle). Less than 5 is to the left (shade left).

19. $i = 12(5) = 60$ inches.

20. The height of a parallelogram is the perpendicular distance from the base to the opposite side, not the slanted side.

21. Base area $= 10 \times 4 = 40$. Height $= 120 \div 40 = 3$ cm.

22. Length $= |3 - (-5)| = 8$. Width $= |4 - (-2)| = 6$. Perimeter $= 2(8) + 2(6) = 28$ units.

23. Base $= 9 - 1 = 8$. Height $= 7 - 1 = 6$. Area $= \frac{1}{2} \times 8 \times 6 = 24$ square units.

24. $6s^2 = 216$, so $s^2 = 36$, giving $s = 6$ cm.

25. Most values are between 22 and 27. The value 45 is far from the rest, making it an outlier.

26. Sum $= 119$. Mean $= 119 \div 7 = 17$. Median (4th of 7) $= 17$. They are equal because the data is symmetric.

Find more at
ViewMath.com/GA-Grade6

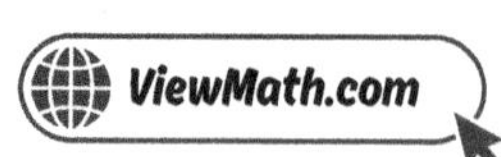

27 Adding 5 to every value shifts the max and min by the same amount, so the difference (range) stays unchanged.

28 A long right whisker means some data values extend far above Q3, indicating the data is skewed to the right.

29 Period 3 has a higher center (both mean and median are higher) and a smaller spread (both range and IQR are smaller). This means Period 3 scored higher on average and their scores were more tightly clustered. Period 1's larger range (40) and IQR (16) suggest greater variability, possibly with some very low or very high scores.

30 $P(5) = \dfrac{1}{6} \approx 0.1\overline{6}$, which rounds to 0.17.

☑ Practice Test 5 — Answer Key

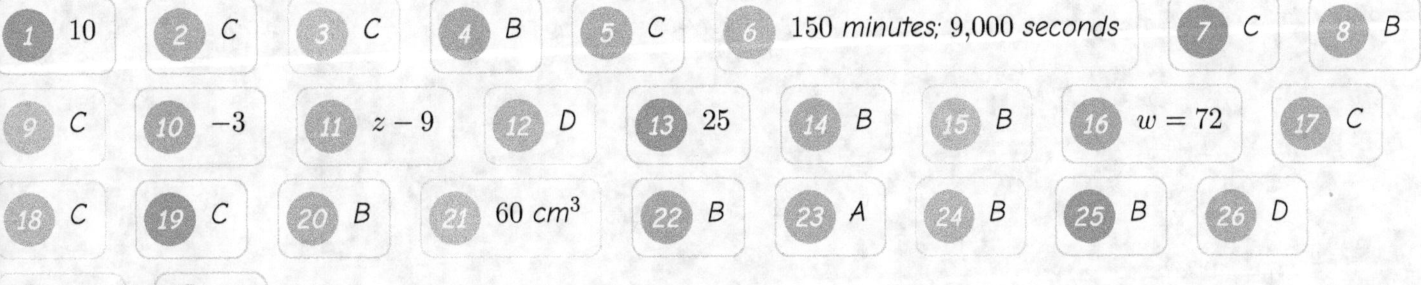

1 10	**2** C	**3** C	**4** B	**5** C	**6** 150 minutes; 9,000 seconds	**7** C	**8** B
9 C	**10** −3	**11** $z - 9$	**12** D	**13** 25	**14** B	**15** B	**16** $w = 72$	**17** C
18 C	**19** C	**20** B	**21** 60 cm^3	**22** B	**23** A	**24** B	**25** B	**26** D
27 2.4	**28** B						

29 Server A median = \$45, Server B median = \$45. Same typical earnings. Server B is more consistent (range = \$10 vs. \$2

30 B

💡 *Time to Learn!* 💡

*Review the explanations below, **especially for the questions you missed**.*

Understanding why each answer is correct builds stronger problem-solving skills.

***Tip:** Circle any questions you got wrong, then read their explanation carefully.*

📖 *Practice Test 5 — Detailed Explanations*

1. *Total parts $= 3 + 2 = 5$. Each part $= 25 \div 5 = 5$. Swimming $= 2 \times 5 = 10$.*

2. *$240 \div 8 = 30$ miles per gallon.*

3. *$4 \times 2.5 = 10$ loaves, so $6 \times 2.5 = 15$ cups. Alternatively, unit rate: $6 \div 4 = 1.5$ cups per loaf, then $1.5 \times 10 = 15$.*

4. *Ratio: $1 : 4$. When $x = 3$: $y = 3 \times 4 = 12$.*

5. *$8 \times 5 = 40$ km.*

6. *$2 \times 60 + 30 = 150$ minutes. $150 \times 60 = 9{,}000$ seconds.*

7. *$\dfrac{2}{3} \times \dfrac{5}{4} = \dfrac{10}{12} = \dfrac{5}{6}$.*

8. *The four steps of the standard algorithm are **Divide, Multiply, Subtract, Bring Down**. After dividing, you multiply the partial quotient by the divisor before subtracting.*

9. *A gain of yards is a positive change. Debt, below sea level, and below zero are all represented by negative numbers.*

Find more at
ViewMath.com/GA-Grade6

10 The distance from 0 to -5 is 5 units in the negative direction. $\frac{3}{5}$ of that distance is $\frac{3}{5} \times 5 = 3$ units. Moving 3 units left from 0 gives -3.

11 "9 fewer than z" means subtract 9 from z: $z - 9$.

12 $6(n + 2)$ means $6 \times (n + 2)$. The two factors being multiplied are 6 and $(n + 2)$.

13 $4^2 + 2(4) + 1 = 16 + 8 + 1 = 25$.

14 Distribute: $4 \times 3y - 4 \times 2 = 12y - 8$.

15 Notebooks cost $3n$ and pens cost $1 \times p = p$. Total: $3n + p$.

16 Multiply by 8: $w = 9 \times 8 = 72$.

17 $t \leq 30$ means 30 or less, which is "no more than 30 degrees."

18 Closed circle means 3 is included ($\geq$ or $\leq$). Shading right means greater values: $x \geq 3$.

19 You choose x (independent), and y is computed from it (dependent). y increases as x increases.

20 When both bases are equal ($b_1 = b_2 = 4$), the trapezoid is actually a parallelogram. Area $= \frac{1}{2}(4 + 4)(6) = 24 = 4 \times 6$.

21 $V = 2.5 \times 4 \times 6 = 60$ cm^3.

22 Horizontal side: $|7 - 1| = 6$. Vertical side: $|2 - (-4)| = 6$. Both sides are 6 units, so it is a square. The side length is 6.

Find more at
ViewMath.com/GA-Grade6

23 Length $= |6 - (-3)| = 9$. Width $= 54 \div 9 = 6$ units.

24 $SA = 2(6)(4) + 2(6)(3) + 2(4)(3) = 48 + 36 + 24 = 108$ cm^2.

25 The values 6 and 7 appear most frequently, and most of the data falls in the 6–7 range.

26 Mean $=$ sum $\div 5 = 20$, so sum $= 20 \times 5 = 100$.

27 Distances from 10: $4, 2, 0, 2, 4$. MAD $= (4 + 2 + 0 + 2 + 4) \div 5 = 12 \div 5 = 2.4$.

28 If the median is closer to Q1, the distance from median to Q3 is larger, meaning the upper half of the middle 50% is more spread out.

29 Server A data: $35, 40, 40, 45, 45, 45, 50, 50, 55, 60$. Median $= (45 + 45)/2 = 45$. Range $= 25$. Server B data: $40, 40, 40, 45, 45, 45, 45, 50, 50, 50$. Median $= (45 + 45)/2 = 45$. Range $= 10$. Same median but Server B is more consistent.

30 $\frac{2}{5} = 0.4$ and $\frac{3}{4} = 0.75$. Since $0.75 > 0.4$, Event B is more likely.

☑ Practice Test 6 — Answer Key

1 Bananas to yogurt: $3 : 4$;   Yogurt to bananas: $4 : 3$ **2** B   **3** B   **4** A   **5** B   **6** B

7 C   **8** B   **9** B   **10** B   **11** C   **12** C   **13** 16   **14** B   **15** C   **16** A

17 $w < 50$ **18** Solutions: 0 and 5 (or any values greater than -1). Non-solution: -1 (or any value ≤ -1)

19 B   **20** 48 in^2 **21** B   **22** 12 units   **23** B   **24** C   **25** A   **26** B   **27** D

Find more at
ViewMath.com/GA-Grade6

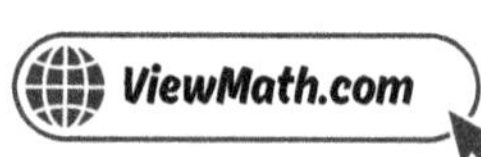

 **28** B    **29** B    **30** B

💡 Time to Learn! 💡

Review the explanations below, **especially for the questions you missed**.

Understanding why each answer is correct builds stronger problem-solving skills.

Tip: Circle any questions you got wrong, then read their explanation carefully.

📖 Practice Test 6 — Detailed Explanations

1 "3 bananas for every 4 cups of yogurt" gives $3 : 4$. Flip the order for yogurt to bananas: $4 : 3$.

2 Brand X: $\$8.75 \div 5 = \$1.75$ each. Brand Y: $\$4.50 \div 3 = \$1.50$ each. Brand Y is cheaper.

3 $2 \times 8 = 16$ students, so $3 \times 8 = 24$ pencils.

4 Hours go on the x-axis, miles on the y-axis: $(1, 30)$, $(2, 60)$, $(3, 90)$.

5 "How much for one" is a unit rate problem. Divide to find the amount per 1 unit.

6 $1 \ kg = 1{,}000 \ g$. $7{,}000 \div 1{,}000 = 7 \ kg$.

7 $4 \div \dfrac{2}{3} = \dfrac{4}{1} \times \dfrac{3}{2} = \dfrac{12}{2} = 6$ batches.

8 After $36 \div 12 = 3$, bringing down 1 gives $1 \div 12 = 0$. The student skipped that zero and jumped to $12 \div 12 = 1$, writing 31 instead of 301.

Find more at
ViewMath.com/GA-Grade6

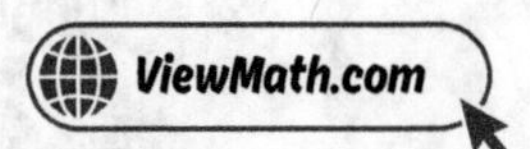

9 A negative temperature means below zero. $-12°F$ means 12 degrees below zero.

10 On a number line, numbers increase from left to right. The correct order from least to greatest is $-1.2 < -0.5 < 0.3 < 1.5$. Choice D shows greatest to least.

11 Half of n is $\frac{n}{2}$. "6 more" means add 6: $\frac{n}{2} + 6$.

12 Part C points to 3, which is a term with no variable — a constant.

13 $8^2 = 64$. Then $64 \div 4 = 16$.

14 Distribute: $3x + 12 + 2x$. Combine: $3x + 2x = 5x$. Result: $5x + 12$.

15 Hourly earnings: $8h$. Plus tip: $8h + 15$.

16 Subtract 3.5: $x = 10 - 3.5 = 6.5$.

17 "Less than 50" means strictly under 50: $w < 50$.

18 Any number greater than -1 is a solution. -1 itself is not, since $>$ does not include the boundary.

19 The independent variable (number of cars) goes on the x-axis.

20 Each triangle: $\frac{1}{2} \times 6 \times 8 = 24$ in^2. Two triangles: $24 \times 2 = 48$ in^2.

21 $V = l \times w \times h$. If h becomes $2h$, then $V_{new} = l \times w \times 2h = 2(lwh)$, so the volume doubles.

22 $|8 - (-4)| = |12| = 12$ units.

Find more at
ViewMath.com/GA-Grade6

ViewMath.com

23 Width $= 56 \div 8 = 7$ units.

24 A rectangular prism has 6 faces: 3 pairs of opposite rectangles.

25 The outlier 100 increases the mean significantly. Without it, the center is around 17. With it, the mean jumps to about 27.

26 The outlier 50 increases the mean to about 15.1, which is higher than 6 of the 7 values. The median (10) better represents a typical value.

27 Range $= 100 - 1 = 99$. The outlier 100 makes the range very large.

28 Lower half: $3, 5, 7$. The median of the lower half is 5, so $Q1 = 5$.

29 The median and IQR are resistant to outliers and work better for skewed data. The mean and MAD are better for symmetric data.

30 Probability always ranges from 0 (impossible) to 1 (certain). It can never be negative or greater than 1.

📋 Practice Test 7 — Answer Key

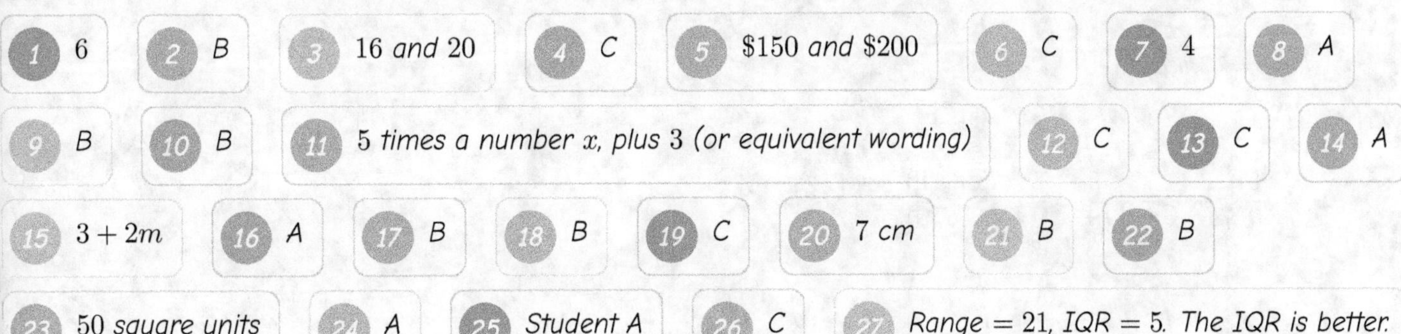

1 6   2 B   3 16 and 20   4 C   5 $150 and $200   6 C   7 4   8 A

9 B   10 B   11 5 times a number x, plus 3 (or equivalent wording)   12 C   13 C   14 A

15 $3 + 2m$ 16 A   17 B   18 B   19 C   20 7 cm   21 B   22 B

23 50 square units   24 A   25 Student A   26 C   27 Range $= 21$, IQR $= 5$. The IQR is better.

Find more at
ViewMath.com/GA-Grade6

 B    B    D

💡 Time to Learn! 💡

*Review the explanations below, **especially for the questions you missed.***

Understanding why each answer is correct builds stronger problem-solving skills.

Tip: *Circle any questions you got wrong, then read their explanation carefully.*

📖 Practice Test 7 — Detailed Explanations

1 $21 \div 7 = 3$, so multiply both by 3: iced tea $= 2 \times 3 = 6$.

2 $150 \div 5 = 30$ gallons per hour.

3 Row 2: $5 \times 2 = 10$, so $8 \times 2 = 16$. Row 3: $8 \times 4 = 32$, so $5 \times 4 = 20$.

4 $5 : 15$ simplifies to $1 : 3$.

5 Total parts $= 7$. Each part $= \$350 \div 7 = \$50$. Family 1: $3 \times \$50 = \$150$. Family 2: $4 \times \$50 = \$200$.

6 1 foot $= 12$ inches. $5 \times 12 = 60$ inches.

7 $2\frac{1}{2} = \frac{5}{2}$. Then $\frac{5}{2} \div \frac{5}{8} = \frac{5}{2} \times \frac{8}{5} = \frac{40}{10} = 4$ shelves.

8 $47 \div 12 = 3$ R11; bring down 5 → $115 \div 12 = 9$ R7; bring down 2 → $72 \div 12 = 6$. Answer: 396. Check: $396 \times 12 = 4{,}752$.

 Find more at
ViewMath.com/GA-Grade6

9. Below sea level is represented by a negative number. 400 feet below sea level is -400.

10. $\frac{11}{4} = 2.75$. Since $2 < 2.75 < 3$, it is between 2 and 3.

11. $5x$ means "5 times x" and $+3$ means "plus 3."

12. $5m - 2n + 7 + m$ has 4 terms: $5m$, $2n$, 7, and m.

13. $2(9) + 2(4) = 18 + 8 = 26$.

14. Like terms: $5p + 2p + 4p = 11p$. Constants: $3 - 1 = 2$. Result: $11p + 2$.

15. Flat fee: \$3. Per-mile cost: $2m$. Total: $3 + 2m$.

16. Divide both sides by 5: $n = 45 \div 5 = 9$.

17. "Greater than 7" uses the $>$ symbol: $x > 7$.

18. $x > 2$ does NOT include 2, so the circle should be open, not closed.

19. Time (hours) is independent — it passes regardless. Battery level depends on time.

20. $84 = \frac{1}{2}(10 + 14)h = \frac{1}{2}(24)h = 12h$. So $h = 84 \div 12 = 7$ cm.

21. $V = 3 \times 2 \times 2 = 12$ ft^3.

22. For a horizontal side (same y), subtract the x-coordinates and take the absolute value.

Find more at
ViewMath.com/GA-Grade6

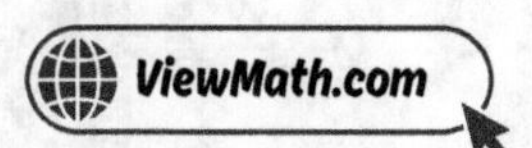

23 Length $= |6 - (-4)| = 10$. Width $= |2 - (-3)| = 5$. Area $= 10 \times 5 = 50$ square units.

24 Net A is a valid cross-shaped cube net. Net B has two squares on the same side, which causes overlap. Net C forms a 2×3 block, which doesn't fold into a cube.

25 A smaller range means the data values are closer together. Student A's data is more consistent because the values span only 5 units.

26 Total $= 15 \times 6 = 90$. Remove 9: $90 - 9 = 81$. New mean $= 81 \div 5 = 16.2$.

27 Range $= 30 - 9 = 21$. Q1: median of $9, 10, 10 = 10$. Q3: median of $14, 15, 30 = 15$. IQR $= 15 - 10 = 5$. The \$30 wage is an outlier that makes the range misleadingly large. The IQR (5) better represents how spread out most wages are.

28 The five-number summary consists of minimum, first quartile (Q1), median, third quartile (Q3), and maximum.

29 The range is 12 and the data is evenly spaced. Whether "very spread out" is accurate depends on context, but the data is quite orderly and not wildly spread.

30 $P(yellow) = \dfrac{2}{8} = \dfrac{1}{4} = 0.25$.

✅ Practice Test 8 — Answer Key

1 B   2 Part A: 0.5 laps per minute (or 1 lap every 2 minutes).   Part B: 7 laps.   3 15   4 B

5 40 minutes   6 C   7 B   8 C   9 B   10 $-0.75,\ -\dfrac{2}{3},\ -\dfrac{1}{2},\ \dfrac{1}{6}$ 11 B   12 B

13 \$23   14 $6x + 12$ 15 $n + 12 - 5$ or $n + 7$ 16 $a = 7$

Get Online

Find more at
ViewMath.com/GA-Grade6

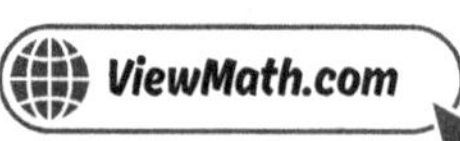

17 $n = 4$ is NOT a solution to $n > 4$ (since 4 is not greater than 4). $n = 4$ IS a solution to $n \geq 4$ (since 4 equals 4).

18 C

19 $w = 45 - 3t$; empty when $t = 15$ minutes

20 B

21 $1 \ ft^3$

22 B

23 C

24 $348 \ in^2$

25 Skewed to the left (or. most data on the right with a tail to the left).

26 C

27 6

28 C

29 Class A performed better (higher median) and was more consistent (smaller IQR).

30 B

💡 Time to Learn! 💡

Review the explanations below, **especially for the questions you missed**.

Understanding why each answer is correct builds stronger problem-solving skills.

Tip: Circle any questions you got wrong, then read their explanation carefully.

📖 Practice Test 8 — Detailed Explanations

1 Dogs have 4 parts. Each part $= 2$ animals. Dogs $= 4 \times 2 = 8$.

2 Part A: From the graph, the swimmer completes 1 lap every 2 minutes, so the rate is $\frac{1}{2}$ lap per minute. Part B: $14 \div 2 = 7$ laps.

3 $2 \times 5 = 10$ strawberries, so $3 \times 5 = 15$ cups of yogurt.

4 x doubled from 4 to 8, so y doubles from 7 to 14.

5 $1,000 \div 250 = 4$ batches. $10 \times 4 = 40$ minutes.

6 1 yard $= 3$ feet. $7 \times 3 = 21$ feet.

Find more at
ViewMath.com/GA-Grade6

7 $\dfrac{2}{7} \times \dfrac{6}{5} = \dfrac{12}{35}$. Kai is correct.

8 $73 \div 24 = 3$ R1; bring down $4 \to 14 \div 24 = 0$ R14; bring down $4 \to 144 \div 24 = 6$. Answer: 306. The zero in the tens place must not be skipped. Check: $306 \times 24 = 7{,}344$.

9 Below sea level is negative. Sea level is 0, so 75 feet below is -75.

10 Convert to decimals: $-\dfrac{2}{3} \approx -0.667$, -0.75, $\dfrac{1}{6} \approx 0.167$, $-\dfrac{1}{2} = -0.5$. From least to greatest: $-0.75 < -0.667 < -0.5 < 0.167$.

11 "8 less than a number" means subtract 8 from n: $n - 8$.

12 The term d means $1 \cdot d$, so the coefficient is 1. (The subtraction sign belongs to the operation, not the coefficient of d itself in the original expression.)

13 $5 + 3(6) = 5 + 18 = 23$ dollars.

14 Perimeter $= 2(2x + 5) + 2(x + 1) = 4x + 10 + 2x + 2 = 6x + 12$.

15 Start with n, add 12, subtract 5: $n + 12 - 5 = n + 7$.

16 $a = 105 \div 15 = 7$.

17 $>$ does not include equality. $\geq$ includes equality.

18 $-3 \geq -3$ is true (they are equal). The other values are less than -3.

19 Set $w = 0$: $0 = 45 - 3t$, so $3t = 45$, $t = 15$ minutes.

Find more at
ViewMath.com/GA-Grade6

20   The student used the slant side instead of the height. Area $= 8 \times 5 = 40$ square units.

21 $V = \frac{1}{3} \times \frac{1}{2} \times 6 = \frac{6}{6} = 1$ ft^3.

22   Same y-coordinate, so it is a horizontal distance. $|6 - (-4)| = |10| = 10$ units.

23   Length $= |5 - (-3)| = 8$. Width $= |2 - (-4)| = 6$. Area $= 8 \times 6 = 48$ square units.

24 $SA = 2(15)(6) + 2(15)(4) + 2(6)(4) = 180 + 120 + 48 = 348$ in^2.

25   The peak is at 5, with values trailing to the left toward 1. Most data is on the higher end.

26   Most values are between 2 and 5, but the outlier 100 increases the sum significantly, pulling the mean up to 22.8.

27   Distances from 60: $10, 5, 0, 5, 10$. $MAD = (10 + 5 + 0 + 5 + 10) \div 5 = 30 \div 5 = 6$.

28   A larger IQR means the middle 50% of the data covers a wider range. Class B's middle scores are more spread out.

29   Class A's median is 8 points higher. Class A's IQR is 12 points smaller, meaning its middle 50% is more tightly clustered.

30   Numbers ≤ 3: $1, 2, 3$. That is 3 out of 6. $P(\leq 3) = \frac{3}{6} = \frac{1}{2}$.

✓ *Practice Test 9 — Answer Key*

1 C **2** B **3** B **4** B **5** C **6** C **7** D **8** 1 **9** D

10 Part A: $A = -\frac{7}{4}$, $B = \frac{1}{2}$, $C = -\frac{1}{4}$; Part B: $-\frac{7}{4}$, $-\frac{1}{4}$, $\frac{1}{2}$ **11** B **12** B **13** A **14** A

15 C **16** C **17** C **18** B **19** C **20** C **21** B **22** C **23** A **24** C

25 C **26** B **27** C **28** B **29** B **30** $\frac{1}{5}$

💡 Time to Learn! 💡

Review the explanations below, **especially for the questions you missed**.

Understanding why each answer is correct builds stronger problem-solving skills.

Tip: Circle any questions you got wrong, then read their explanation carefully.

📖 Practice Test 9 — Detailed Explanations

1 Total parts $= 1 + 3 + 2 = 6$. Each part $= 18 \div 6 = 3$. Blue $= 3 \times 3 = 9$.

2 Divide: $120 \div 4 = 30$ pages per minute.

3 $3 \times 2 = 6$, so $5 \times 2 = 10$.

4 $1 : 3$ multiplied by 3 gives $3 : 9$, so $(3, 9)$ is on the line.

5 $3 \times 4 = 12$ raisins, so $4 \times 4 = 16$ cups of granola.

6 $1\,L = 1,000\,mL$. $2.5 \times 1,000 = 2,500\,mL$.

7 $\dfrac{4}{5} \div \dfrac{2}{5} = \dfrac{4}{5} \times \dfrac{5}{2} = \dfrac{20}{10} = 2$ servings.

8 After subtracting 15, the remainder is 0. Bring down 7 to get 7. Since $7 \div 5 = 1$ R2, the missing digit in the quotient is 1. The full answer is 315.

9 A withdrawal removes money, creating a negative change. -15 represents losing or going below by 15. A deposit, a temperature above zero, and an elevation above sea level are all positive.

10 Part A: The number line is divided into fourths. Point A is one-fourth to the right of -2, so $A = -2 + \dfrac{1}{4} = -\dfrac{7}{4}$. Point B is two-fourths (one-half) to the right of 0, so $B = \dfrac{1}{2}$. Point C is one-fourth to the left of 0, so $C = -\dfrac{1}{4}$.
Part B: $-\dfrac{7}{4} < -\dfrac{1}{4} < \dfrac{1}{2}$, that is $-1.75 < -0.25 < 0.5$.

11 The product of 4 and t is $4t$. The sum of 9 and that product is $9 + 4t$.

12 The terms are $4x$, 9, and $2y$. Terms are separated by $+$ or $-$ signs.

13 $6(4) - 4^2 = 24 - 16 = 8$.

14 $7n$ and $4n$ are like terms. Add coefficients: $7 + 4 = 11$, so $7n + 4n = 11n$.

15 s represents the side length of any square. It can be any positive number.

16 Multiply by 6: $t = 5 \times 6 = 30$.

17 $5 > 5$ is false. 5 equals 5, but is not greater than 5. It would be a solution to $x \geq 5$.

18 Open circle means -2 is NOT included ($<$ or $>$). Shading left means less than: $x < -2$.

Find more at
ViewMath.com/GA-Grade6

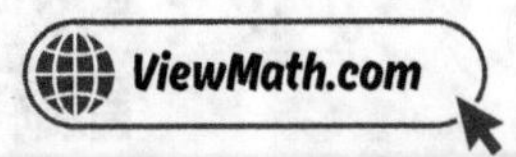

19 $c = 7(6) = 42$ dollars.

20 Area $= 25 \times 14 = 350 \ m^2$. Bags needed: $350 \div 50 = 7$ bags.

21 Box A volume $= 6 \times 4 \times 5 = 120$. Box B: $120 = 10 \times 3 \times h = 30h$, so $h = 4$.

22 Same x-coordinate, so it is a vertical distance. $|5 - (-3)| = |8| = 8$ units.

23 Base $= 10$. Height $= 4$. Area $= \frac{1}{2} \times 10 \times 4 = 20$ square units.

24 A cube has 6 equal faces. SA $= 6 \times 5^2 = 6 \times 25 = 150 \ in^2$.

25 The dot plot shows a peak at 13 with roughly equal numbers of dots on each side. This is approximately symmetric.

26 Original mean: $(58 + 60 + 62 + 64) \div 4 = 61$. New mean: $(58 + 60 + 62 + 64 + 82) \div 5 = 326 \div 5 = 65.2$.

27 $IQR = Q3 - Q1 = 55 - 35 = 20$.

28 Even number of values. The two middle values are 6 and 8. Median $= (6 + 8) \div 2 = 7$.

29 The median (72) is the typical value. The IQR (15) tells you the middle half of the data spans 15 points.

30 Multiples of 5 from 1 to 20: $5, 10, 15, 20$. That is 4 favorable outcomes out of 20. $P = \dfrac{4}{20} = \dfrac{1}{5}$.

✅ Practice Test 10 — Answer Key

Find more at
ViewMath.com/GA-Grade6

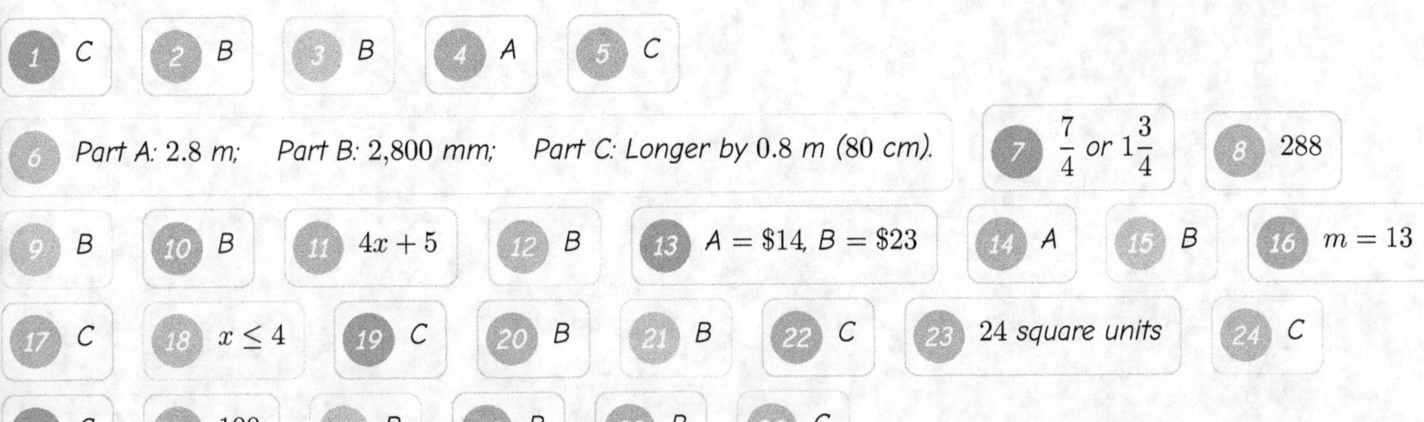

| 1 | C | 2 | B | 3 | B | 4 | A | 5 | C |

6. Part A: 2.8 m;   Part B: 2,800 mm;   Part C: Longer by 0.8 m (80 cm).

7. $\frac{7}{4}$ or $1\frac{3}{4}$

8. 288

9. B

10. B

11. $4x + 5$

12. B

13. $A = \$14, B = \$23$

14. A

15. B

16. $m = 13$

17. C

18. $x \leq 4$

19. C

20. B

21. B

22. C

23. 24 square units

24. C

25. C

26. 100

27. B

28. B

29. B

30. C

💡 Time to Learn! 💡

Review the explanations below, **especially for the questions you missed**.

Understanding why each answer is correct builds stronger problem-solving skills.

Tip: Circle any questions you got wrong, then read their explanation carefully.

📖 Practice Test 10 — Detailed Explanations

1. Apple has 3 parts, orange has 5 parts. Total parts = 8. Total ounces = $8 \times 4 = 32$.

2. Store A: $\$4.50 \div 10 = \$0.45$. Store B: $\$3.20 \div 8 = \$0.40$. Store B is cheaper per pencil.

3. $2 : 3$ multiplied by 3 gives $6 : 9$. The other pairs do not simplify to $2 : 3$.

4. The ratio is $2 : 3$. Choice A continues with $6 : 9 = 2 : 3$. Choice B has $6 : 8$ which is not $2 : 3$.

5. $42 \div 6 = 7$ sections, which needs $7 + 1 = 8$ posts (one at each end). Note: if counting "per 6 feet" as a rate, $42 \div 6 = 7$, but fences need a post at the start too, giving 8.

Find more at
ViewMath.com/GA-Grade6

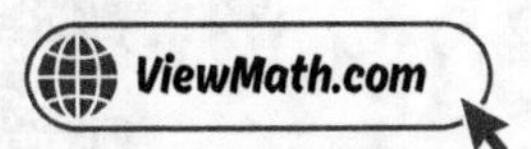

(6) *Part A: $280 \div 100 = 2.8$ m. Part B: $280 \times 10 = 2,800$ mm. Part C: $2.8 - 2 = 0.8$ m longer.*

(7) $\dfrac{7}{10} \times \dfrac{5}{2} = \dfrac{35}{20} = \dfrac{7}{4} = 1\dfrac{3}{4}.$

(8) *$60 \div 21 = 2$ R18; bring down $4 \to 184 \div 21 = 8$ R16; bring down $8 \to 168 \div 21 = 8$. Answer: 288. Check: $288 \times 21 = 6,048.$*

(9) *On a horizontal number line, negative numbers are always to the left of zero and positive numbers are to the right.*

(10) *$-\dfrac{5}{3} \approx -1.667$. Since $-2 < -1.667 < -1$, it is between -2 and -1. A common mistake is placing it between -1 and 0 by confusing $\dfrac{5}{3}$ with $\dfrac{2}{3}$.*

(11) *There are 4 triangles (each worth x) and 5 unit squares. The expression is $4x + 5$.*

(12) *$\dfrac{n}{5} = \dfrac{1}{5} \times n$. The factors are $\dfrac{1}{5}$ and n.*

(13) *$h = 2$: $8 + 3(2) = 14$. $h = 5$: $8 + 3(5) = 23$.*

(14) *Distribute: $7k - 21 + 5$. Combine: $-21 + 5 = -16$. Result: $7k - 16$.*

(15) *In 4 days you read $4r$ pages. Remaining: $100 - 4r$.*

(16) *Subtract 17: $m = 30 - 17 = 13$.*

(17) *"More than \$25" means greater than 25: $d > 25$.*

(18) *Closed circle at 4 (included) and shading to the left (less than): $x \le 4$.*

Find more at
ViewMath.com/GA-Grade6

ViewMath.com

19 Each dog has 4 legs: $L = 4d$.

20 $A = \frac{1}{2}(5 + 13)(8) = \frac{1}{2}(18)(8) = 72\ ft^2$.

21 $V = 5 \times 3 \times 4 = 60\ cm^3$.

22 The base goes from $(0, 0)$ to $(8, 0)$. Distance $= |8 - 0| = 8$ units.

23 Base $= |5 - (-3)| = 8$. Height $= |4 - (-2)| = 6$. Area $= \frac{1}{2} \times 8 \times 6 = 24$ square units.

24 The shaded face is $8\ cm \times 5\ cm = 40\ cm^2$.

25 Most data values are between 8 and 12, clustering around 10. The 25 is an outlier. A typical time is about 10 minutes.

26 Need total $= 90 \times 6 = 540$. Current total $= 85 + 90 + 78 + 92 + 95 = 440$. Need $540 - 440 = 100$.

27 Lower half: $4, 6, 8$. The median of the lower half is 6, so $Q1 = 6$.

28 $Q1 = 30$, $Q3 = 70$. IQR $= 70 - 30 = 40$.

29 Little overlap means most values in one group are different from the values in the other group, which signals a clear difference between the groups.

30 Total tokens: $6 + 4 + 2 = 12$. $P(blue) = \frac{6}{12} = \frac{1}{2}$.

Well done checking your answers!

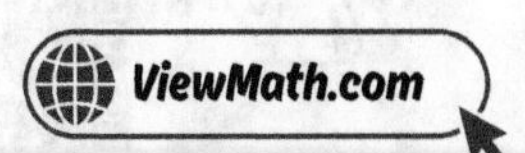

Keep practicing to strengthen your skills.

Author's Final Note

I hope you enjoyed this book as much as I enjoyed writing it. Whether you are a student working through the material, a parent supporting your child's learning, or a teacher guiding your class, I have tried to make this book as clear and engaging as possible. I hope I have succeeded. If you have any suggestions for improvement, please let me know. I would love to hear from you.

The accuracy of calculations is very important to me. We have done our best, but I also expect that I have made some minor errors. Constant improvement is the name of the game. If you find any errors, please let me know. I will fix them in the next edition.

For students: Your learning journey does not end here. I have written a series of books to help you learn math. Make sure you browse through them. I especially recommend workbooks and practice tests to help you prepare for your exams.

For parents: Thank you for investing in your child's education. I encourage you to explore the companion resources available online to help support your child outside the classroom.

For teachers: Thank you for the invaluable work you do every day. I hope this book serves as a useful resource in your classroom. Feel free to reach out if you have suggestions or would like to discuss how best to use this book with your students.

I also enjoy reading your reviews. If you have a moment, please leave a review on where you found this book. It will help others find this book. If you have any questions or comments, please feel free to contact me at drNazari@ViewMath.com.

And one last thing: Remember to use online resources for additional help. I recommend using the resources on `https://ViewMath.com` You can find video lessons, practice problems, and more. You can also use the online companion for this book to track your progress and access additional resources.

Wishing all students the best in their studies, parents every success in supporting their children, and teachers continued inspiration in their classrooms!

Dr. A. Nazari